High-Temperature Matrix Cracking, Opening and Closure in Ceramic-Matrix Composites

Ceramic-matrix composites (CMCs) can withstand higher temperatures, reduce cooling airflow, improve turbine efficiency, and greatly reduce structural mass compared to high-temperature alloys. This book focuses on matrix first/ multiple cracking and crack opening and closure behavior in CMCs at high temperatures.

While conducting *in situ* experimental observations to analyze the damage mechanisms and failure modes, the author develops micromechanical damage models and constitutive models to predict the first matrix cracking stress, multiple matrix cracking density, matrix crack opening displacement, and cracking closure stress at high temperatures. The effects of the composite's constituent properties, stress level, and ambient temperature on matrix cracking and opening and closure behavior are also discussed.

This book will help material scientists and engineering designers to understand and master matrix cracking and closure behavior of fiber-reinforced CMCs.

Longbiao Li is a Lecturer at the College of Civil Aviation, Nanjing University of Aeronautics and Astronautics. His research focuses on the vibration, fatigue, damage, fracture, reliability, safety, and durability of aircraft and aero engines. In this research area, he is the first author of 204 SCI journal publications (59 JCR Q1), 11 monographs, 4 edited books, 4 textbooks, 3 book chapters, 33 Chinese patents, 2 US patents, 2 Chinese software copyrights, and more than 30 refereed conference proceedings. He has been involved in different projects related to structural damage, reliability, and airworthiness design for aircraft and aero engines, supported by the Natural Science Foundation of China, COMAC Company, and AECC Commercial Aircraft Engine Company.

High-Temperature Matrix Cracking, Opening and Closure in Ceramic-Matrix Composites

Longbiao Li

CRC Press
Taylor & Francis Group
Boca Raton London New York

CRC Press is an imprint of the
Taylor & Francis Group, an **informa** business

Designed cover image: © frank_peters

First edition published 2024
by CRC Press
2385 NW Executive Center Drive, Suite 320, Boca Raton FL 33431

and by CRC Press
4 Park Square, Milton Park, Abingdon, Oxon, OX14 4RN

CRC Press is an imprint of Taylor & Francis Group, LLC

© 2024 Longbiao Li

ISBN: 978-1-032-63751-8 (hbk)
ISBN: 978-1-032-63857-7 (pbk)
ISBN: 978-1-032-63850-8 (ebk)

DOI: 10.1201/9781032638508

Typeset in Minion
by MPS Limited, Dehradun

To My Son Shengning Li

Contents

Preface

H IGH-PERFORMANCE AERO ENGINES ARE THE BASIS FOR THE DEVELOPMENT OF advanced military and civil aircraft. The operating temperature of aero engines can reach 1200–1650°C, and the life requirement is more than 1000 h. The materials used in aero engines should possess good performance stability during service and exhibit a metal-like fracture behavior, e.g., insensitive to cracks and no catastrophic damage. Therefore, there is an urgent need to develop a new generation of high-temperature long-life thermal structure materials. Ceramic-matrix composites (CMCs) can withstand higher temperatures, reduce cooling airflow, improve turbine efficiency, and greatly reduce structural mass compared to high-temperature alloys. Continuous fiber-reinforced silicon carbide CMCs (SiC-CMC including C/SiC and SiC/SiC) have already been applied to aero engine hot-section components, e.g., regulating/sealing plates, combustion liners, turbine guide vanes, and turbine outer rings.

Under mechanical loading, multiple damage mechanisms occur and contribute to the nonlinear mechanical behavior of CMCs, e.g., matrix cracking, interface debonding, and fibers fracture and pullout. This book focuses on matrix first/multiple cracking and crack opening and closure behavior in CMCs at high temperatures. *In situ* experimental observations were conducted to analyze the damage mechanisms and failure modes. Micromechanical damage models and constitutive models were developed to predict the first matrix cracking stress, multiple matrix cracking density, matrix crack opening displacement, and cracking closure stress at high temperatures. Effects of composite's constituent properties, stress level, environmental temperature on matrix cracking, and opening and closure behavior are discussed.

This book covers the damage mechanisms, damage models, micro-mechanical constitutive model for matrix cracking, and opening and closure behavior of CMCs. I hope this book helps material scientists and engineering designers to understand and master matrix cracking and closure behavior of fiber-reinforced CMCs.

Longbiao Li

CHAPTER 1

Introduction

FABRICATION METHODS OF CMCS

Continuous fiber-reinforced ceramic-matrix composites (CMCs) retain the advantages of ceramic materials such as high-temperature resistance, oxidation resistance, wear resistance, and corrosion resistance, while overcoming the inherent defects of low fracture toughness and poor resistance to external impact loads of ceramic materials. CMCs usually consist of three parts: reinforcing fibers, interface layer, and ceramic matrix, and their properties are determined by the properties of each part and their interaction [1].

CMCs have the following advantages:

- Lightweight. CMCs have low density (1/3–1/4 of high-temperature alloys) and can be used in combustion chambers, regulators/seals, and other components to directly reduce mass by about 50%.

- High-temperature resistance. CMCs can operate at temperatures up to 1650°C, simplifying or even eliminating the cooling structure, optimizing the engine structure, and increasing the engine operating temperature and service life. It can be used at 1200°C for a long time without cooling structure.

- Excellent high-temperature oxidation resistance performance. CMCs are able to maintain high stability in high-temperature oxidative environments, reducing the development and application costs of thermal protective coatings.

- Superior mechanical properties. The mechanical properties of CMCs are qualitatively improved relative to monolithic ceramics through preparation process optimization, especially the interfacial layer design.

DOI: 10.1201/9781032638508-1

1

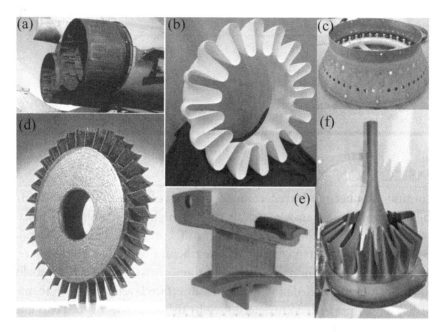

FIGURE 1.1 Typical CMC components applied in aero engines: (a) regulating/sealing sheets of M88-2 aero engine; (b) CMC mixer on the AE3007 engine; (c) CMC combustion liner of CFM56 aero engine; (d) Turbine blisk developed by NASA; (e) Turbine guide vane developed by NASA; and (f) CMC center cone and mixer of CFM56-5C aero engine.

CMCs have been successfully applied to aero engine components [2,3], as shown in Figure 1.1. CMC regulating/sealing sheets produced by SAFRAN have been used for more than 10 years on the M88-2 engine of the Rafale and the M53 engine of the Mirage 2000 fighter jet, and its designed CMC tail nozzle was certified by the European Aviation Safety Agency for commercial flight use on April 22, 2015, and completed its first commercial flight on June 16, 2015, on a CFM56-5B engine. Rolls-Royce (UK), in cooperation with NASA, has adopted the CMC mixer on the AE3007 engine to reduce engine NO_x emissions and noise. SAFRAN's CMC combustion chamber liners were tested and validated on the CFM56 engine, reducing combustion chamber cooling gas consumption by 35%, and its CMC mixer and center cone were used on the CFM56-5C engine, which was flight tested and validated on the Airbus A320 in February 2012. NASA prepared and tested CMC turbine guide vanes and turbine blisk in a real engine environment in the Ultra-Efficient Engine Technology program with a 15–25% reduction in turbine cooling air consumption and a turbine inlet temperature of over 1650°C. General Electric tested CMC combustion chamber and high-pressure turbine components in the GEnx engine ground test. The components using CMC include the inner and outer rings of the combustion chamber flame tube, the high-pressure turbine first-stage cowling ring, and the second-stage guide vane.

Chemical Vapor Infiltration

The chemical vapor infiltration (CVI) fabrication process consists of CVI of a fiber coating and a ceramic matrix into a 2D (e.g., stacked fabric) or 3D fiber architecture. During this process, the CVI reagents are pumped into a furnace containing a heated preform. These gaseous reagents infiltrate the preform and react at the surface of the fibers, initially building up an interface (fiber coating), followed by deposition of matrix materials, thereby densifying the preform into a CMC that typically contains 10–20% porosity. The CVI process contains extremely complex thermodynamic and kinetic processes and requires control of many process parameters. The advantages of CVI process are less fiber damage during preparation, high purity and crystalline integrity of the prepared ceramic matrix, and high mechanical properties of the composites. However, its disadvantages are equally obvious, long manufacturing cycle time, high cost, and high porosity of the prepared composites.

Polymer Infiltration and Pyrolysis

Polymer infiltration and pyrolysis (PIP) process uses polymer liquid-phase precursors as impregnates to obtain densified composites through multi-cycle cross-linking curing and high-temperature pyrolysis. This process was first used for the preparation of C/C composites with bitumen or resin polymer precursors and has gradually been extended to the preparation of CMCs with a high degree of technological maturity. The advantages of this process include the ability to produce single-phase or complex-phase ceramics with controlled components and structure through the design of precursor components; lower pyrolysis temperature, reducing the damage to the fibers during heat treatment; the ability to achieve near-net molding, reducing post-processing costs; and the ability to produce large components with complex shapes. The disadvantages are the polymer precursor pyrolysis process is accompanied by large volume shrinkage, which causes some damage to the fibers; the single-cycle ceramic yield is low, which needs to be processed by multi-cycle impregnation pyrolysis; and the manufacturing cycle is long. The prepared CMCs possess high porosity.

Melt Infiltration

The preparation of SiC/SiC composites using melt infiltration (MI) involves four main steps:

- Deposition of coatings (interphase) on SiC fiber surfaces using the CVI process.

- Continued deposition of a certain thickness (about 3–5 μm) of SiC matrix on the surface of the fiber coating.

- Introducing the slurry casting.

- Liquid-phase MI (or reactive melt infiltration (RMI)) of silicon or silicon alloys.

Depending on the presence or absence of carbon and silicon reactions during the melting process, there are two types of melting: RMI and non-reactive MI. This process is simple, short cycle time, and produces composites with very high densities and porosity below 2%.

MECHANICAL PROPERTIES OF CMCS

In this section, the mechanical properties of mini CVI-SiC/SiC, uni-directional MI-SiC/SiC, cross-ply MI SiC/SiC, and 2D woven CVI-SiC/SiC composites were analyzed.

Mini CVI-SiC/SiC Composites

Figure 1.2 shows the tensile stress–strain curve and tangent modulus (E_t) versus strain (ε_c) curve of a mini CVI-SiC/SiC composite [3]. For the mini-SiC/SiC composite with multilayered $(BN/SiC)_4$ interphase, the composite's elastic modulus E_c upon initial loading was $E_c = 197 \pm 34$ GPa. The value of the first matrix cracking stress (FMCS) was $\sigma_{mc} = 422 \pm 14$ MPa with the corresponding strain of $\varepsilon_{mc} = 0.30 \pm 0.04\%$. The composite's UTS was $\sigma_{uts} = 863 \pm 42$ MPa and the fracture strain was $\varepsilon_f = 0.87 \pm 0.05\%$. The composite's tangent modulus E_t decreased with increase in strain, and the initial composite tangent modulus range was between $E_t = 141$ and 231 GPa, and before final fracture, the composite tangent modulus range was between $E_t = 87$ and 106 GPa.

Unidirectional MI-SiC/SiC Composites

Figure 1.3 shows the experimental tensile stress–strain curve and tangent modulus (E_t) versus strain (ε_c) curve for a unidirectional MI-SiC/SiC

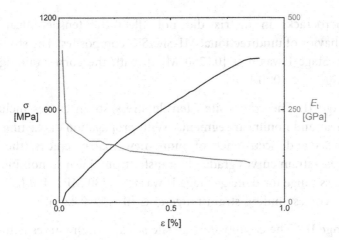

FIGURE 1.2 Tensile stress–strain curve and tangent modulus versus strain curve of mini CVI-SiC/SiC composites.

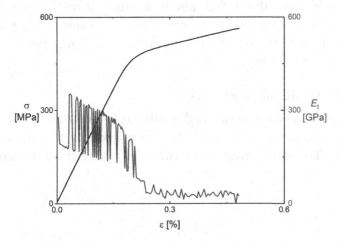

FIGURE 1.3 Tensile stress–strain curve and tangent modulus versus strain curve of unidirectional MI-SiC/SiC composites.

composite [4]. The average composite's elastic modulus (E_c) was approximately E_c = 249 GPa; the average composite's proportional limit stress (σ_{pls}) was approximately σ_{pls} = 452 MPa. The average tensile strength (σ_{uts}) was approximately σ_{uts} = 567 MPa, with an average fracture strain (ε_f) of approximately ε_f = 0.43%. The tensile stress–strain curve showed obvious nonlinearity, revealing three stages:

- **Stage I**. The composite's tensile stress–strain curves exhibited a linear elastic pattern, and micro-matrix may occur. However, these microcracks in matrix did not affect the tensile linear elastic behavior of unidirectional MI-SiC/SiC composites. The stress range for Stage I was $\sigma \in [0, 280\ \text{MPa}]$, with the corresponding strain range $\varepsilon_c \in [0, 0.11\%]$.

- **Stage II**. The composite's tensile stress–strain curves include the linear and nonlinear segments. With propagation of existing matrix cracks and occurrence of more new matrix cracks, the tensile stress–strain curves gradually transfer from linear to nonlinear. The stress range for damage Stage II was $\sigma \in [280\ \text{MPa}, 480\ \text{MPa}]$, with the corresponding strain range $\varepsilon_c \in [0.11\%, 0.22\%]$.

- **Stage III**. The composite's tensile stress–strain curves exhibited a secondary quasi-linear elastic pattern accompanied by saturated matrix cracking and gradual fibers fracture. Due to the high tensile strength of SiC fibers, gradual fibers fracture mainly occurs at Stage III. The stress range for damage Stage III was $\sigma \in [480\ \text{MPa}, 562\ \text{MPa}]$, with the corresponding strain range $\varepsilon_c \in [0.22\%, 0.48\%]$.

Cross-Ply MI-SiC/SiC Composites

Figure 1.4 shows the experimental tensile stress–strain curve and tangent modulus (E_t) versus strain (ε_c) curve for a cross-ply MI-SiC/SiC composite [4]. The tensile stress–strain curves in unidirectional and cross-ply

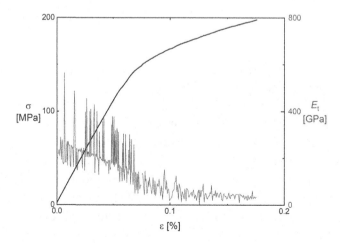

FIGURE 1.4 Tensile stress–strain curve and tangent modulus versus strain curve of cross-ply MI-SiC/SiC composites.

MT-SiC/SiC composites exhibit the same trend but have large variations in applied stress and strain values, mainly due to the damage mechanisms of transverse cracking in 90° plies and low ECFL ($\chi = 0.5$) for cross-ply laminates. The average composite's elastic modulus E_c was approximately 228 GPa, and average composite's proportional limit stress σ_{pls} was approximately $\sigma_{pls} = 129$ MPa. Average composite's ultimate tensile strength σ_{uts} was 182 MPa, with an average fracture strain (ε_f) of approximately $\varepsilon_f = 0.14\%$. The tensile stress–strain curve showed obvious nonlinearity, revealing three stages:

- **Stage I**. The composite's tensile stress–strain curves exhibited a linear elastic pattern, and transverse cracking in the 90° plies may occur. However, these transverse cracks in 90° plies did not affect the tensile linear elastic behavior of cross-ply MI-SiC/SiC composites. The stress range for Stage I was $\sigma \in [0, 40 \text{ MPa}]$, with the corresponding strain range $\varepsilon_c \in [0, 0.017\%]$.

- **Stage II**. The composite's tensile stress–strain curves include the linear and nonlinear segments. With propagation of existing matrix cracks and occurrence of more new matrix cracks, the tensile stress–strain curves gradually transfer from linear to nonlinear. The stress range for damage Stage II was $\sigma \in [40 \text{ MPa}, 160 \text{ MPa}]$, with the corresponding strain range $\varepsilon_c \in [0.017\%, 0.089\%]$.

- **Stage III**. The composite's tensile stress–strain curves exhibited a secondary quasi-linear elastic pattern accompanied by saturated matrix cracking and gradual fibers fracture. Due to the high tensile strength of SiC fibers, gradual fibers fracture mainly occurs at Stage III. The stress range for damage Stage II was $\sigma \in [160 \text{ MPa}, 197.8 \text{ MPa}]$, with the corresponding strain range $\varepsilon_c \in [0.089\%, 0.176\%]$.

2D Woven CVI-SiC/SiC Composites

Figure 1.5 shows the monotonic tensile curve and related composite's tangent modulus (E_t) evolution with strain (ε_c) curve of a 2D plain-woven CVI-SiC/SiC composite at an elevated temperature of $T = 900°C$ in air atmosphere [5]. The tensile curve at 900°C exhibits nonlinear damage behavior due to internal multiple damage mechanisms of matrix cracking, interface debonding, and fiber failure. Through the analysis of the tensile curve, the composite's initial elastic modulus was $E_c = 122$ GPa, the composite's proportional limit stress was $\sigma_{pls} = 128$ MPa with the corresponding

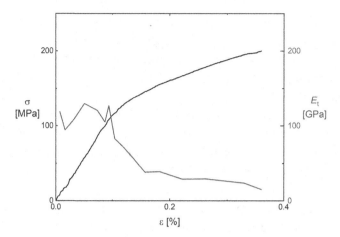

FIGURE 1.5 (a) Tensile curve and (b) tangent modulus versus strain curve of 2D plain-woven CVI-SiC/SiC composites at an elevated temperature of $T = 900°C$ in air atmosphere.

strain $\varepsilon_{pls} = 0.12\%$, the composite's saturation matrix cracking stress was $\sigma_{sat} = 145$ MPa with the corresponding strain $\varepsilon_{smc} = 0.16\%$, and the composite's tensile strength was $\sigma_{uts} = 200$ MPa with the fracture strain $\varepsilon_f = 0.36\%$. The composite's tensile curve and related tangent modulus can be divided into three stages, including:

- Stage I, the linear elastic stage with the high composite's tangent modulus. The corresponding stress range for Stage I is between the initial loading stress and the proportional limit stress (i.e., $\sigma_{pls} = 128.3$ MPa).

- Stage II, the nonlinear damage stage due to multiple damage mechanisms (e.g., matrix cracking, interface debonding, and fiber failure) with the rapid degradation of composite's tangent modulus. The corresponding stress range for Stage II is between the proportional limit stress (i.e., $\sigma_{pls} = 128.3$ MPa) and matrix cracking saturation stress (i.e., $\sigma_{sat} = 145$ MPa).

- Stage III, the secondary linear damage stage with slow degradation of composite's tangent modulus, due to the gradual fibers broken with increasing applied stress. The corresponding stress range for Stage III is between the matrix cracking saturation stress (i.e., $\sigma_{sat} = 145$ MPa) and composite's tensile strength (i.e., $\sigma_{uts} = 200$ MPa).

DESIGN AND TESTING OF CMC COMPONENTS

SiC/SiC Turbine Twin Guide Vanes

The SiC/SiC high-pressure turbine twin guide vanes were fabricated using the CVI method [6]. The twin guide vanes were formed by 2D lamination. Figure 1.6 shows the photograph of the SiC/SiC twin guide vanes observed under X-ray computed tomography (XCT). The thermal shock experiments of CVI SiC/SiC high-pressure turbine guide vanes were conducted at $T = 1400$, 1450, and 1480°C with heating time $t = 30$ s, holding time $t = 30$ s, and cooling time $t = 60$ s. Figure 1.7 shows the figure of the thermal shock test for turbine guide vanes. After cyclic thermal shock tests, the damage regions were observed under a scanning electronic microscope.

Figure 1.8 shows the surface temperature distribution for the twin guide vanes under thermal shock test at the target temperature of $T = 1400$°C. The temperature in the middle of the leading edge of the left guide vane was the highest. The temperature at the basin region decreased from the leading edge to the trailing edge, and the temperature of the trailing edge was the lowest. The temperature at the trailing edge and the leading edge of the right guide vane was the highest, and the temperature of the middle region was the lowest. The temperature of the basin region was higher than that of the back region of the guide vane.

Before the thermal shock test, the surface of turbine guide vanes was smooth. When the thermal shock test temperature increased to 1480°C and the holding time increased from 30 to 60 s, delamination phenomenon appeared on the upper surface of the guide vane near the trailing edge, as shown in Figure 1.9.

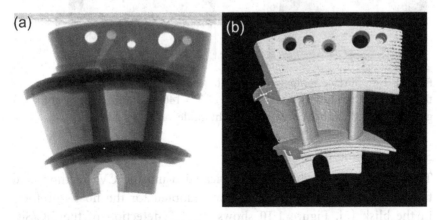

FIGURE 1.6 Photograph of SiC/SiC twin guide vanes observed under XCT: (a) front view and (b) back view.

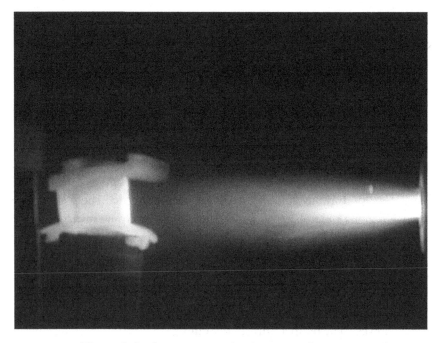

FIGURE 1.7   Thermal shock experiment for SiC/SiC turbine twin guide vanes.

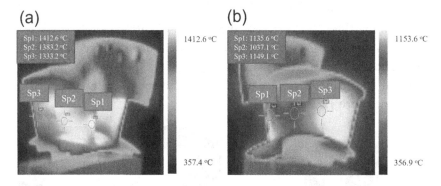

FIGURE 1.8   Surface temperature distribution for the twin guide vanes under thermal shock test at target temperature of $T = 1400°C$: (a) basin area of the left guide vane and (b) back area of the right guide vane.

SiC/SiC Turbine Blisk

The SiC/SiC turbine blisk was fabricated using the CVI method, and the Spider Web Structure (SWS) was adopted for the fiber's preform in the blisk [5]. Figure 1.10 shows the CT detection of the SiC/SiC turbine blisk.

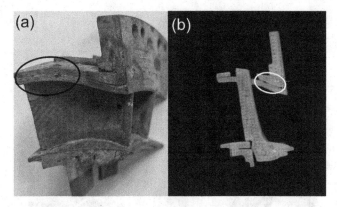

FIGURE 1.9 Thermal shock test of SiC/SiC twin guide vanes at target temperature $T = 1480°C$ at $N = 200$: (a) delamination occurred at the trailing edge and (b) delamination observed under XCT.

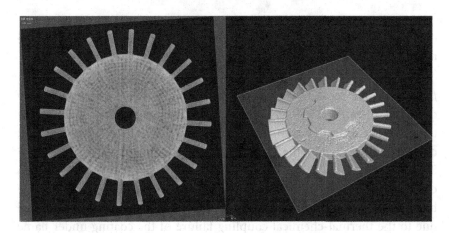

FIGURE 1.10 SiC/SiC turbine blisk detected under the CT.

According to the load spectrum, the micro-turbine engine with the SWS-SiC/SiC turbine blisk was tested. The target rotating speed is set to be $n = 85,000$ r/min. After reaching the specified target speed, the blisk is increased by 5000 r/min each time until the designed target speed is 112,000 r/min. After each test, the natural frequency of the turbine blisk is measured by laser vibration meter, and the internal damage state in the disk was qualitatively analyzed by the change of the natural frequency. When the maximum rotation speed increases from $n_{max} = 85,000$ r/min to 105,000 r/min, the first-order natural frequency decreases from

FIGURE 1.11   Coating spalling of the SWS-SiC/SiC turbine blisk during the rotating test.

f = 14,426 Hz to 13,681 Hz, and the second-order natural frequency decreases from f = 15,394 Hz to 14,188 Hz. Experimental results show that the first-order natural frequency of the SWS-SiC/SiC turbine blisk decreased by 5% with the increase in the rotating speed from n_{max} = 85,000 r/min to n_{max} = 105,000 r/min, which indicates that there is damage in the disk, which leads to the decrease in the stiffness of the blisk. During the rotating test, multiple coating spalling was observed due to the thermal-chemical coupling failure of the coating under flame impingement, as shown in Figure 1.11.

MICROSTRUCTURE AND DAMAGE MECHANISMS OF CMCS

Under tensile loading, multiple damage mechanisms occur in CMCs. Figure 1.12 shows the experimental and predicted tensile stress–strain curves, interface debonding ratio (η), and AE energy versus loading time curves of a 2D SiC/SiC composite. Under tensile loading, the composite exhibits obviously nonlinear behavior due to multiple damage mechanisms of matrix cracking, interface debonding, and fibers broken. Based on the monitoring of acoustic emission, the damage mechanisms under tensile loading are analyzed as follows:

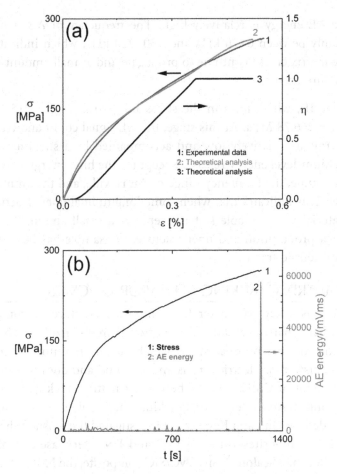

FIGURE 1.12 (a) Experimental and predicted tensile stress–strain and interface debonding ratio curves; (b) the stress and acoustic energy versus loading time curves of 2D SiC/SiC composites.

1. Stage I, which starts from initial loading to applied stress σ_{mc} = 50 MPa. At this stage, the AE count and AE energy are in the middle level, the duration and amplitude are basically stable after a period of rising range, and the frequency range of AE is wide, which indicates that the specimen gradually produces damage in the initial stage of loading, but the energy of single damage is small, mainly due to initial micro-cracking and partial interface debonding.

2. Stage II, which lies in the stress range from σ_{mc} = 50 MPa to σ_{sat} = 215 MPa. At this stage, the AE count is in the middle level, and

the AE energy is relatively high. The frequency of AE signals lies mainly between 0–70 kHz and 130–220 kHz, which indicates that the matrix cracks continue to propagate and a small amount of fiber fracture occurs.

3. Stage III, which lines in the stress range from σ_{sat} = 215 MPa to σ_{uts} = 266.78 MPa. At this stage, the AE signal count and AE signal energy are relatively low and accompanied by a small amount of medium-level energy release. Except for the high energy release in a short time, the frequency range of AE is wide, and the amplitude is low. This indicates that when some longitudinal fibers fracture, the material is in a stable state; except for a small amount of matrix crack propagation and fiber fracture, the sample has little damage until tensile fracture.

FIRST MATRIX CRACKING BEHAVIOR OF CMCS

The FMCS is a key parameter for composite structure or component design. Micro-matrix cracking first occurred in the matrix-rich region due to the thermal residual stress which can be monitored using the acoustic emission or electrical resistance method and does not affect the linear behavior of CMCs. When these short matrix cracking propagates into the long steady-state matrix cracking, the tensile stress–strain curve begins to deflect. Guo and Kagawa [7] investigated the tensile behavior of 2D SiC/SiC composites with the PyC and BN interphase at high temperature. For the Nicalon^{TM} SiC/PyC/SiC composite, the FMCS decreased from σ_{mc} = 65 MPa at T = 298 K to σ_{mc} = 33 MPa at T = 1200 K; and for the Hi-Nicalon^{TM} SiC/PyC/SiC composite, the FMCS decreased from σ_{mc} = 75 MPa at T = 298 K to σ_{mc} = 45 MPa at T = 1400 K. Experimental and predicted FMCS versus temperature curves of Nicalon^{TM} SiC/PyC/SiC and Hi-Nicalon^{TM} SiC/PyC/SiC composites are shown in Figure 1.13.

MULTIPLE MATRIX CRACKING EVOLUTION OF CMCS

At high temperatures, the generation and propagation of matrix cracking consume the energy inside CMCs, which slows down or prevents further matrix cracking propagation and achieves the toughness behavior. Temperature dependence on composite's constituent properties affects the evolution of matrix multiple cracking. The density and openings of matrix cracks depend on the fiber architecture, fiber/matrix interface bonding intensity, applied load, and environments. Figure 1.14 shows

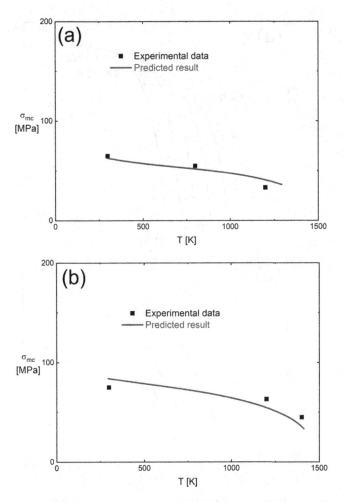

FIGURE 1.13   Experimental and predicted FMCS versus the temperature curves of (a) SiC/SiC composites with the PyC interphase and (b) SiC/SiC composites with the BN interphase.

the experimental and theory predicted matrix cracking density of SiC/SiC composites at room temperature and high temperatures of $T = 773$, 873, 973, and 1073 K. At room temperature, the matrix multiple cracking evolution occurs from $\sigma_{mc} = 240$ MPa with matrix cracking density $\lambda_m = 1.1/\text{mm}$ to saturation at $\sigma_{sat} = 320$ MPa with $\lambda_m = 13/\text{mm}$. At high temperature $T = 1073$ K, the matrix cracking density increases from $\lambda_m = 0.35/\text{mm}$ at $\sigma_{mc} = 169$ MPa to $\lambda_m = 9.4/\text{mm}$ at $\sigma_{sat} = 236$ MPa, and the interface debonding ratio increases from $\eta = 0.8$ to 66.9%.

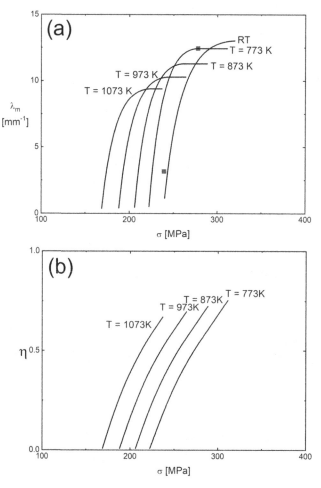

FIGURE 1.14 (a) Experimental and theoretical matrix cracking density versus applied stress curves; (b) the fiber/matrix interface debonding ratio versus applied stress curves of SiC/SiC composites.

MATRIX CRACK OPENING BEHAVIOR OF CMCS

In the stress environment, matrix cracks occur first. The appearance and opening of cracks increase the oxygen diffusion channel and accelerate the oxidation failure of CMCs. Figure 1.15 shows the experimental and predicted crack opening displacement (u_{cod}), crack opening stress (σ_{cos}), interface debonding ratio (η), interface complete debonding stress (σ_{icds}), and axial fiber's and matrix's displacements (w_f and w_m) for different matrix crack lengths.

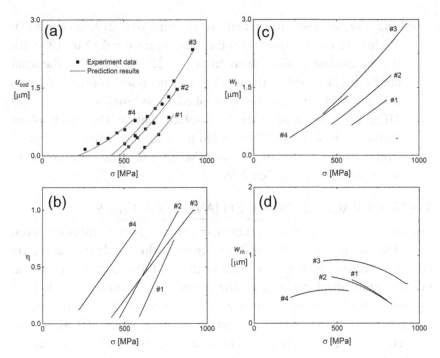

FIGURE 1.15 (a) u_{cod}, (b) η, (c) w_f, and (d) w_m of SiC/SiC minicomposites at room temperature.

- For matrix crack #1, the crack opening stress was approximately $\sigma_{cos} = 592$ MPa; the experimental u_{cod} increased from $u_{cod} = 0.18$ to 0.85 μm with the increase in applied stress from $\sigma = 627$ to 774 MPa, and the predicted u_{cod} increased from $u_{cod} = 0.03$ μm to 0.88 μm with the increase in applied stress from $\sigma = 592$ to 802 MPa; the interface debonding ratio increased from $\eta = 0.06$ to 0.73 with the increase in applied stress from $\sigma = 592$ to 802 MPa; the axial fiber's displacement at the crack plane increased from $w_f = 0.67$ μm to 1.22 μm with the increase in applied stress from $\sigma = 592$ to 802 MPa; and the axial matrix's displacement at the crack plane decreased from $w_m = 0.64$ μm to 0.33 μm with the increase in applied stress from $\sigma = 592$ to 802 MPa.

- For matrix crack #4, the crack opening stress was approximately $\sigma_{cos} = 219$ MPa; the experimental u_{cod} increased from $u_{cod} = 0.14$ to 0.78 μm with the increase in applied stress from $\sigma = 260$ to 569 MPa, and the predicted u_{cod} increased from $u_{cod} = 0.008$ to 0.83 μm with the increase in applied stress from $\sigma = 219$ to 569 MPa; the interface debonding ratio increased from $\eta = 0.12$ to 0.83 with the increase in applied stress from $\sigma = 219$ to 569 MPa; the axial fiber's displacement at the crack plane increased from $w_f = 0.39$ to 1.32 µm with the increase in applied stress from $\sigma = 219$ to 569 MPa; and the axial matrix's displacement at the crack plane increased from $w_m = 0.38$ to 0.496 µm with the increase in applied stress from $\sigma = 219$ to 469 MPa, and then decreased to $w_m = 0.48$ µm at $\sigma = 569$ MPa.

MATRIX CRACK CLOSURE BEHAVIOR OF CMCS

For some CMCs, the matrix cracks can contact during unloading, even when the matrix is subject to residual tension, which is defined as matrix cracking closure. Such behavior arises due to the occurrence of lateral grain-to-grain displacements as the matrix cracks form. The thermal expansion coefficient of the fiber and the matrix changes with temperature, indicating the change of thermal residual stress of CMCs with temperature. Figures 1.16–1.18 show the experimental and predicted mechanical hysteresis loops and damage parameter of interface reverse slip ratio (ϕ_u) of a mini-SiC/SiC composite under $\sigma_{max} = 890$, 935, and 1078 MPa. As the matrix cracking density changes with the increase in applied stress upon loading and decrease in applied stress upon unloading, the unloading and reloading strain coincide with each other before unloading to the valley stress. When the tensile peak stress increases from $\sigma_{max} = 890$ to 1078 MPa, the composite's unloading

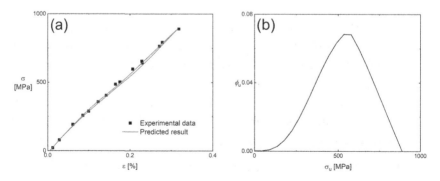

FIGURE 1.16 (a) Hysteresis loops and (b) ϕ_u versus σ_u curve under $\sigma_{max} = 890$ MPa of mini-SiC/SiC composites.

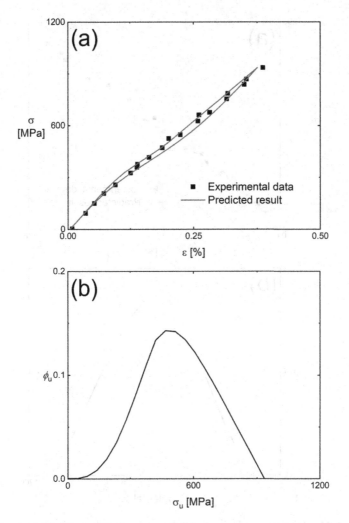

FIGURE 1.17 (a) Hysteresis loops and (b) ϕ_u versus σ_u curve under σ_{max} = 935 MPa of mini-SiC/SiC composites.

residual strain increases a little, due to the existence of thermal residual stress in the SiC matrix. Upon unloading, the damage parameter ϕ_u changes with the decrease in applied stress. ϕ_u increased from $\phi_u = 0$ to the peak value and then decreases to $\phi_u = 0$. The value of CCS can be determined from the curve of ϕ_u versus σ_u curves. The CCS is $\sigma_{ccs} = 108$ MPa for the mini-SiC/SiC composite.

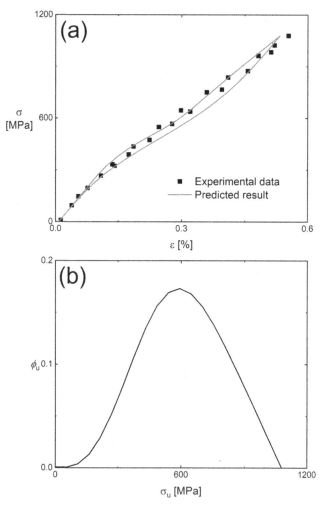

FIGURE 1.18 (a) Hysteresis loops and (b) ϕ_u versus σ_u curve under σ_{max} = 1078 MPa of mini-SiC/SiC composites.

SUMMARY AND CONCLUSIONS

This chapter introduced the fabrication methods, mechanical properties, design and testing of typical components, microstructure, and damage mechanisms of CMCs. The mechanical behavior of first matrix cracking, multiple matrix cracking evolution, matrix crack opening, and matrix crack closure in CMCs were also introduced.

REFERENCES

1. Li LB. *High-temperature mechanical hysteresis behavior of ceramic-matrix composites*. CRC Press, Taylor & Francis Group, London, UK, 2022.
2. Li LB. *High temperature mechanical behavior of ceramic-matrix composites*. Wiley-VCH, Weinheim, Germany, 2021.
3. Lü X, Li LB, Sun J, Yang J, Jiao J. Microstructure and tensile behavior of (BN/SiC)$_n$ coated SiC fibers and SiC/SiC minicomposites. *J. Eur. Ceram. Soc.* 2023; 43:1828–1842.
4. Liu H, Li LB, Yang JH, Zhou YR, Ai Y, Qi Z, Gao Y, Jiao J. Characterization and modeling damage and fracture of prepreg-MI SiC/SiC composite under tensile loading at room temperature. *Appl. Compos. Mater.* 2022; 29:1167–1193.
5. Guo X, Li J, Zeng Y, Huang X, Li LB, Xu YL, Hu X. Design, fabrication and testing of CVI-SiC/SiC turbine blisk under different load spectrums at elevated temperature. *High Temp. Mater. Process.* 2022; 41:279–288.
6. Liu X, Guo X, Xu Y, Li LB, Zhu W, Zeng Y, Li J, Luo X, Hu X. Cyclic thermal shock damage behavior in CVI SiC/SiC high-pressure turbine twin guide vanes. *Materials* 2021; 14:6104.
7. Guo S, Kagawa Y. Tensile fracture behavior of continuous SiC fiber-reinforced SiC matrix composites at elevated temperatures and correlation to in situ constituent properties. *J. Eur. Ceram. Soc.* 2002; 22:2349–2356.

CHAPTER 2

High-Temperature First Matrix Cracking Behavior in Ceramic-Matrix Composites

INTRODUCTION

Continuous SiC/SiC fiber-reinforced ceramic-matrix composites (CMCs) have excellent properties such as high specific strength, high specific modulus, wear resistance, oxidation resistance, corrosion resistance, radiation resistance, insensitivity to cracks, and non-catastrophic fracture, which makes it a new type of thermal structural material in the application prospects of aviation, aerospace, and energy [1–3]. SiC/SiC composites are mainly used for hot section components of high-performance aero engines and industrial gas turbines and have potential applications in nuclear fusion reactors and fission reactors [4–9]. Solar Turbines Company uses SiC/SiC as the combustion chamber lining of Solar's Centaur 505 engine. In the 35,000 h test run, the NO_x and CO content of the exhaust gas is lower than that of the ordinary engine [10]. A comparative test of fiber-reinforced CMCs combustor lining was carried out in Germany. The results show that after 10 h test in a gas engine, the delamination and debonding between the substrate and the coating appeared in the CVD-SiC-coated C/SiC combustor; however, the SiC/SiC combustor did not suffer any damage after 90 h test [11]. In cooperation with SNECMA to develop fiber-reinforced CMC nozzle

DOI: 10.1201/9781032638508-2

seals for F100-PW-229 aero engine, P&W is also using CMC nozzle flaps and seals validated under the IHPTET program to improve the F119 aero engine, which powered the world's most advanced fighter F-22. With the new flaps, the durability of the aero engine is improved significantly, while the quality and the cost are reduced. GE has signed a multi-year development contract with Goodrich to develop C/SiC nozzle flaps and seals for the higher-temperature F414 engine. Goodrich is responsible for providing lightweight and long-life CMCs, and GE is responsible for testing and evaluation. Now, GE has conducted production and flight testing of CMC standard parts. GE also developed and demonstrated the CMC combustion chamber with the support of the TECH56 program. The CMC combustion chamber can provide high-temperature rise and possess long life and need few cooling airs.

The nonlinear stress–strain behavior of fiber-reinforced CMCs under tensile loading is mainly due to the internal damages of matrix cracking and the fiber/matrix interface debonding [12]. The macro tensile curves of fiber-reinforced CMCs can be divided into three stages, including (1) the linear elastic stage; (2) the nonlinear stage of matrix cracking propagation and interface debonding stage till the saturation of the matrix cracking; and (3) the fibers failure stage after the saturation of matrix cracking [13–16]. Among the three stages mentioned above, the first matrix cracking stress (FMCS) is a key parameter for composite structure or component design. The theoretical analysis of FMCS can be divided into two cases, i.e., the energy balance approach, including the ACK model [17], BHE model [18], SH model [19], and Chiang model [20–22], and the stress intensity factor approach, including MCE model [23], MC model [24], and Chiang model [25]. Pavia et al. [26] predicted the FMCS in micro/nanohybrid brittle matrix composites based on the ACK shear-lag model. It was found that the presence of a small function of strong stiff nanotubes provides significant enhancements in the FMCS. Acoustic emission and electrical resistance can be used to monitor the matrix cracking behavior in fiber-reinforced CMCs [16]. The micro-matrix cracking first occurred in the matrix-rich region due to the thermal residual stress which can be monitored using the acoustic emission or electrical resistance method and does not affect the linear behavior of fiber-reinforced CMCs. When these short matrix cracking propagates into the long steady-state matrix cracking, the tensile stress–strain curve begins to deflect. Li [27] investigated the interface properties of the evolution of multiple matrix cracking of fiber-reinforced CMCs. Low interface shear stress (ISS) leads to the low matrix cracking density. Singh [28] investigated the effect of ISS on the FMCS of SiC/Zircon composite. With the increase in ISS, the FMCS increases. At elevated temperature, the ISS changes with temperature due to the thermal expansion coefficient mismatch between the fiber and the matrix.

In this chapter, the temperature- and time-dependent FMCS of SiC/SiC composites is investigated using the energy balance approach. Effects of the fiber volume, fiber/matrix interface properties, and matrix properties on the temperature- and time-dependent FMCS and composite internal damages are analyzed. Experimental FMCS and fiber/matrix interface debonding length of 2D SiC/SiC composites with different interphases at elevated temperature are predicted.

TEMPERATURE-DEPENDENT FMCS OF SiC/SiC COMPOSITES

In this section, the temperature-dependent FMCS of SiC/SiC composites is investigated using the energy balance approach. The effect of environment temperature on the fiber and matrix elastic modulus, fiber/matrix ISS and interface debonding energy, and the matrix fracture energy are considered. Experimental FMCS and fiber/matrix interface debonding length of 2D SiC/SiC composites with different interphases at elevated temperature are predicted.

Micromechanical Model

The energy balance relationship to evaluate the FMCS of fiber-reinforced CMCs can be determined by Eq. (2.1) [18]

$$
\begin{aligned}
\frac{1}{2} \int_{-\infty}^{\infty} &\left\{ \frac{V_f}{E_f(T)} [\sigma_{fu}(T) - \sigma_{fd}(T)]^2 + \frac{V_m}{E_m(T)} [\sigma_{mu}(T) - \sigma_{md}(T)]^2 \right\} dx \\
&+ \frac{1}{2\pi R^2 G_m(T)} \int_{-l_d(T)}^{l_d(T)} \int_{r_f}^{R} \left[\frac{r_f \tau_i(x,T)}{r} \right] 2\pi r dr dx \\
&= V_m \Gamma_m(T) + \frac{4 V_f l_d(T)}{r_f} \Gamma_i(T)
\end{aligned}
\tag{2.1}
$$

where V_f and V_m are the fiber and matrix volume fraction, respectively; $E_f(T)$ and $E_m(T)$ are the temperature-dependent fiber and matrix elastic modulus, respectively; and $\Gamma_m(T)$ and $\Gamma_i(T)$ are the temperature-dependent matrix fracture energy and interface debonding energy, respectively.

$$\sigma_{fu}(T) = \frac{E_f(T)}{E_c(T)}\sigma \tag{2.2}$$

$$\sigma_{mu}(T) = \frac{E_m(T)}{E_c(T)}\sigma \tag{2.3}$$

$$\sigma_{fd}(x, T) = \begin{cases} \dfrac{\sigma}{V_f} - \dfrac{2\tau_i(T)}{r_f}x, & x \in [0, l_d(T)] \\[2ex] \dfrac{E_f(T)}{E_c(T)}\sigma, & x \in \left[l_d(T), \dfrac{l_c(T)}{2}\right] \end{cases} \tag{2.4}$$

$$\sigma_{md}(x, T) = \begin{cases} 2\dfrac{V_f}{V_m}\dfrac{\tau_i(T)}{r_f}x, & x \in [0, l_d(T)] \\[2ex] \dfrac{E_m(T)}{E_c(T)}\sigma, & x \in \left[l_d(T), \dfrac{l_c(T)}{2}\right] \end{cases} \tag{2.5}$$

$$l_d(T) = \frac{r_f V_m E_m(T)\sigma}{2V_f E_c(T)\tau_i(T)} - \sqrt{\frac{r_f V_m E_f(T)\Gamma_i(T)}{E_c(T)\tau_i^2(T)}} \tag{2.6}$$

where

$$\tau_i(T) = \tau_0 + \mu\frac{|\alpha_{rf}(T) - \alpha_{rm}(T)|(T_m - T)}{A} \tag{2.7}$$

Substituting the upstream and downstream temperature-dependent fiber and matrix axial stresses in Eqs. (2.2), (2.3), (2.4), and (2.5) and the temperature-dependent fiber/matrix interface debonding length in Eq. (2.6) into Eq. (2.1), the energy balance equation leads to the following equation:

$$a\sigma^2 + b\sigma + c = 0 \tag{2.8}$$

where

$$a = \frac{V_m E_m(T)l_d(T)}{V_f E_f(T)E_c(T)} \tag{2.9}$$

$$b = -\frac{2\tau_i(T)}{r_f E_f(T)} l_d^2(T) \tag{2.10}$$

$$c = \frac{4}{3}\left[\frac{\tau_i(T)}{r_f}\right]^2 \frac{V_f E_c(T)}{V_m E_f(T) E_m(T)} l_d^3(T) - \frac{4V_f \zeta_d(T)}{r_f} - V_m \Gamma_m(T) \tag{2.11}$$

Experimental Comparisons

Guo and Kagawa [29] investigated the tensile behavior of 2D SiC/SiC composites with the PyC and BN interphase at elevated temperature. The experimental tensile stress–strain curves of Nicalon™ SiC/PyC/SiC and Hi-Nicalon™ SiC/BN/SiC composites at room and elevated temperatures are shown in Figures 2.1 and 2.2, respectively. For Nicalon™ SiC/PyC/SiC composites, the FMCS decreases from σ_{mc} = 65 MPa at T = 298 K to σ_{mc} = 33 MPa at T = 1200 K; and for Hi-Nicalon™ SiC/PyC/SiC composites, the FMCS decreases from σ_{mc} = 75 MPa at T = 298 K to σ_{mc} = 45 MPa at T = 1400 K. The experimental and predicted FMCS versus

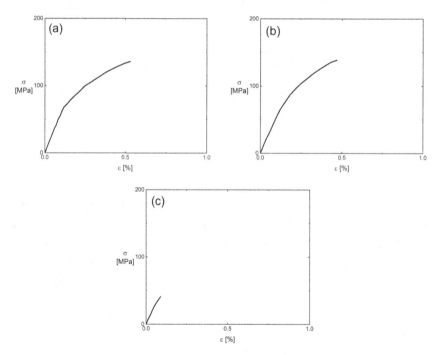

FIGURE 2.1 Experimental tensile stress–strain curves of Nicalon™ SiC/PyC/ SiC at (a) T = 298 K, (b) T = 800 K, and (c) T = 1200 K.

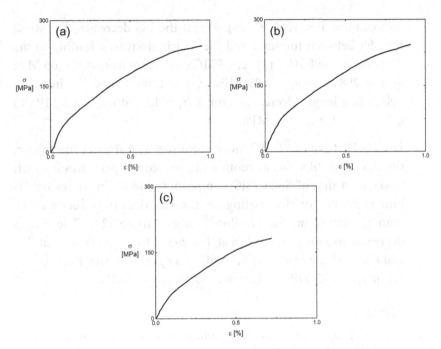

FIGURE 2.2 Experimental tensile stress–strain curves of Nicalon^{TM} SiC/BN/SiC at (a) $T = 298$ K, (b) $T = 1200$ K, and (c) $T = 1400$ K.

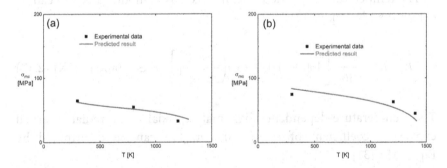

FIGURE 2.3 Experimental and predicted FMCS versus the temperature curves of (a) SiC/SiC composites with the PyC interphase and (b) SiC/SiC composites with the BN interphase.

temperature curves of Nicalon^{TM} SiC/PyC/SiC and Hi-Nicalon^{TM} SiC/PyC/SiC are shown in Figure 2.3.

- For 2D Nicalon^{TM} SiC/SiC composites with the PyC interphase, the ISS of SiC/C/SiC composites decreases at elevated temperature $T = 800$ K from that of $T = 298$ K and then increases again at elevated temperature $T = 1200$ K [29]. When the ISS decreases, the stress transfer between the fiber and the matrix decreases, leading to the decrease in the FMCS [1]. The FMCS decreases from $\sigma_{mc} = 65$ MPa at $T = 298$ K to $\sigma_{mc} = 33$ MPa at $T = 1200$ K; and the interface debonding length decreases from $l_d/r_f = 12.8$ at $\sigma_{mc} = 65$ MPa to $l_d/r_f = 7.3$ at $\sigma_{mc} = 33$ MPa.

- For 2D Hi-Nicalon^{TM} SiC/SiC composites with the BN interphase, the ISS of SiC/BN/SiC at room and elevated temperatures is much lower than that of SiC/C/SiC composites, due to the better oxidation resistance of BN coating on the Hi-Nicalon^{TM} fiber surface than C-coating on the Nicalon^{TM} fiber surface [29]. The FMCS decreases from $\sigma_{mc} = 75$ MPa at $T = 298$ K to $\sigma_{mc} = 45$ MPa at $T = 1400$ K; and the interface debonding length decreases from $l_d/r_f = 7.3$ at $\sigma_{mc} = 75$ MPa to $l_d/r_f = 2.9$ at $\sigma_{mc} = 45$ MPa.

Discussions

The material properties of SiC/SiC composites are given by [30] $V_f = 30\%$, $r_f = 7.5$ μm, $\Gamma_m = 25$ J/m^2 (at room temperature), $\Gamma_i = 0.1$ J/m^2 (at room temperature), $\alpha_{rf} = 2.9 \times 10^{-6}$ /K, and $\alpha_{lf} = 3.9 \times 10^{-6}$/K.

The temperature-dependent SiC matrix elastic modulus $E_m(T)$ can be determined by Eq. (2.12) [31]

$$E_m(T) = \frac{350}{460}\left[460 - 0.04T \exp\left(-\frac{962}{T}\right)\right], \quad T \in [300K\ 1773K] \quad (2.12)$$

The temperature-dependent SiC matrix axial and radial thermal expansion coefficient of $\alpha_{lm}(T)$ and $\alpha_{rm}(T)$ can be determined by Eq. (2.13) [31]

$$\alpha_{lm}(T) = \alpha_{rm}(T) = \begin{cases} -1.8276 + 0.0178T - 1.5544 \times 10^{-5}T^2 \\ \quad + 4.5246 \times 10^{-9}T^3, \quad T \in [125K\ 1273K] \\ 5.0 \times 10^{-6}/K, \quad T > 1273K \end{cases}$$

$$(2.13)$$

The temperature-dependent interface debonding energy $\Gamma_i(T)$ and the matrix fracture energy $\Gamma_m(T)$ can be determined by Eq. (2.14) and (2.15) [32]

$$\Gamma_i(T) = \Gamma_{i0} \left[1 - \frac{\int_{T_0}^{T} C_P(T)\,dT}{\int_{T_0}^{T_m} C_P(T)\,dT} \right] \tag{2.14}$$

$$\Gamma_m(T) = \Gamma_{m0} \left[1 - \frac{\int_{T_0}^{T} C_P(T)\,dT}{\int_{T_0}^{T_m} C_P(T)\,dT} \right] \tag{2.15}$$

where T_0 denotes the reference temperature; T_m denotes the fabricated temperature; Γ_{i0} and Γ_{m0} denote the interface debonding energy and matrix fracture energy at the reference temperature of T_0, respectively.

$$C_P(T) = 76.337 + 109.039 \times 10^{-3}T - 6.535 \times 10^5 T^{-2} - 27.083 \times 10^{-6}T^2 \tag{2.16}$$

Effects of the fiber volume, interface properties, and matrix properties on the temperature-dependent FMCS of SiC/SiC composites are discussed.

Effect of the Fiber Volume on Temperature-Dependent FMCS
The FMCS and the fiber/matrix interface debonding length versus the temperature curves for different fiber volumes (i.e., $V_f = 25\%$, 30%, and 35%) are shown in Figure 2.4. When the temperature increases from $T = 873$ to 1273 K, the FMCS and fiber/matrix interface debonding length decrease with the increase in temperature. At the same temperature, the FMCS increases with the fiber volume, and the interface debonding

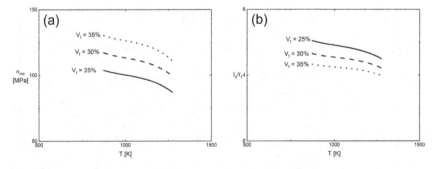

FIGURE 2.4 Effect of the fiber volume on (a) the FMCS versus temperature curves; (b) the fiber/matrix interface debonding length versus temperature curves of SiC/SiC composites.

length decreases with the fiber volume. When the fiber volume increases, the stress transfer between the fiber and the matrix increases, and the stress carried by the matrix increases, leading to an increase in the FMCS and a decrease in the fiber/matrix interface debonding length.

- When the fiber volume is $V_f = 25\%$, the FMCS decreases from $\sigma_{mc} = 103$ MPa at $T = 873$ K to $\sigma_{mc} = 87$ MPa at $T = 1273$ K; and the interface debonding length decreases from $l_d/r_f = 6.1$ at $\sigma_{mc} = 103$ MPa to $l_d/r_f = 5.0$ at $\sigma_{mc} = 87$ MPa.

- When the fiber volume is $V_f = 30\%$, the FMCS decreases from $\sigma_{mc} = 117$ MPa at $T = 873$ K to $\sigma_{mc} = 99$ MPa at $T = 1273$ K; and the interface debonding length decreases from $l_d/r_f = 5.3$ at $\sigma_{mc} = 117$ MPa to $l_d/r_f = 4.4$ at $\sigma_{mc} = 99$ MPa.

- When the fiber volume is $V_f = 35\%$, the FMCS decreases from $\sigma_{mc} = 130$ MPa at $T = 873$ K to $\sigma_{mc} = 110$ MPa at $T = 1273$ K; and the interface debonding length decreases from $l_d/r_f = 4.6$ at $\sigma_{mc} = 130$ MPa to $l_d/r_f = 3.9$ at $\sigma_{mc} = 110$ MPa.

Effect of the ISS on Temperature-Dependent FMCS

The FMCS and the fiber/matrix interface debonding length versus the temperature curves for different ISS (i.e., $\tau_0 = 15$, 20, and 25 MPa) are shown in Figure 2.5. When the temperature increases from $T = 873$ to 1273 K, the FMCS and fiber/matrix interface debonding length decrease with the increase in temperature. The FMCS increases with the ISS, and the interface debonding length decreases with the ISS at the same

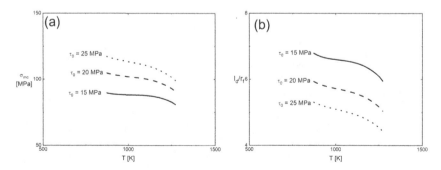

FIGURE 2.5   Effect of the interface shear stress on (a) the FMCS versus temperature curves; (b) the fiber/matrix interface debonding length versus temperature curves of SiC/SiC composites.

temperature. When the ISS increases, the stress transfer between the fiber and the matrix increases, leading to an increase in the FMCS and a decrease in the interface debonding length.

- When the fiber/matrix ISS is τ_0 = 15 MPa, the FMCS decreases from σ_{mc} = 89 MPa at T = 873 K to σ_{mc} = 81 MPa at T = 1273 K; and the interface debonding length decreases from l_d/r_f = 6.8 at σ_{mc} = 89 MPa to l_d/r_f = 5.9 at σ_{mc} = 81 MPa.
- When the fiber/matrix ISS is τ_0 = 20 MPa, the FMCS decreases from σ_{mc} = 104 MPa at T = 873 K to σ_{mc} = 90 MPa at T = 1273 K; and the interface debonding length decreases from l_d/r_f = 5.9 at σ_{mc} = 104 MPa to l_d/r_f = 5 at σ_{mc} = 90 MPa.
- When the fiber/matrix ISS is τ_0 = 25 MPa, the FMCS decreases from σ_{mc} = 117 MPa at T = 873 K to σ_{mc} = 99 MPa at T = 1273 K; and the interface debonding length decreases from l_d/r_f = 5.3 at σ_{mc} = 117 MPa to l_d/r_f = 4.4 at σ_{mc} = 99 MPa.

Effect of the Interface Frictional Coefficient on Temperature-Dependent FMCS

The FMCS and the fiber/matrix interface debonding length versus the temperature curves for different interface frictional coefficients (i.e., μ = 0.02, 0.03, and 0.04) are shown in Figure 2.6. When the temperature increases from T = 873 to 1273 K, the FMCS and interface debonding length decrease with the increase in temperature. At the same temperature, the FMCS decreases with the interface frictional coefficient, and

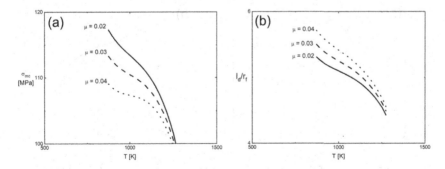

FIGURE 2.6 Effect of the interface frictional coefficient on (a) the FMCS versus temperature curves; (b) the fiber/matrix interface debonding length versus temperature curves of SiC/SiC composites.

the interface debonding length increases with the interface frictional coefficient. When the interface frictional coefficient increases, the ISS decreases at the same temperature, leading to a decrease in the FMCS and an increase in the interface debonding length.

- When the interface frictional coefficient is $\mu = 0.02$, the FMCS decreases from $\sigma_{mc} = 117$ MPa at $T = 873$ K to $\sigma_{mc} = 99$ MPa at $T = 1273$ K; and the interface debonding length decreases from $l_d/r_f = 5.3$ at $\sigma_{mc} = 117$ MPa to $l_d/r_f = 4.4$ at $\sigma_{mc} = 99$ MPa.

- When the interface frictional coefficient is $\mu = 0.03$, the FMCS decreases from $\sigma_{mc} = 113$ MPa at $T = 873$ K to $\sigma_{mc} = 98$ MPa at $T = 1273$ K; and the interface debonding length decreases from $l_d/r_f = 5.5$ at $\sigma_{mc} = 113$ MPa to $l_d/r_f = 4.4$ at $\sigma_{mc} = 98$ MPa.

- When the interface frictional coefficient is $\mu = 0.04$, the FMCS decreases from $\sigma_{mc} = 109$ MPa at $T = 873$ K to $\sigma_{mc} = 97$ MPa at $T = 1273$ K; and the interface debonding length decreases from $l_d/r_f = 5.7$ at $\sigma_{mc} = 109$ MPa to $l_d/r_f = 4.5$ at $\sigma_{mc} = 97$ MPa.

Effect of the Interface Debonding Energy on Temperature-Dependent FMCS

The FMCS and the fiber/matrix interface debonding length versus the temperature curves for different interface debonding energies (i.e., $\Gamma_i = 0.1$, 0.2, and 0.3 J/m^2) are shown in Figure 2.7. When the temperature increases from $T = 873$ to 1273 K, the FMCS and fiber/matrix interface debonding length decrease with the increase in temperature. At the same temperature,

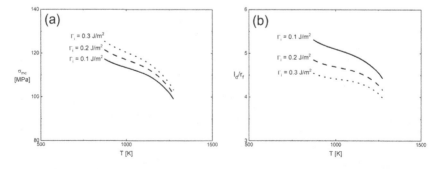

FIGURE 2.7   Effect of the interface debonding energy on (a) the FMCS versus temperature curves; (b) the fiber/matrix interface debonding length versus temperature curves of SiC/SiC composites.

the FMCS increases with the interface debonding energy, and the fiber/matrix interface debonding length decreases with the interface debonding energy. When the interface debonding energy increases, the resistance for the interface debonding increases, leading to the increase in the FMCS and the decrease in the interface debonding length.

- When the interface debonding energy is $\Gamma_i = 0.1$ J/m^2, the FMCS decreases from $\sigma_{mc} = 117$ MPa at $T = 873$ K to $\sigma_{mc} = 99$ MPa at $T = 1273$ K; and the interface debonding length decreases from $l_d/r_f = 5.3$ at $\sigma_{mc} = 117$ MPa to $l_d/r_f = 4.4$ at $\sigma_{mc} = 99$ MPa.

- When the fiber/matrix interface debonding energy is $\Gamma_i = 0.2$ J/m^2, the FMCS decreases from $\sigma_{mc} = 121$ MPa at $T = 873$ K to $\sigma_{mc} = 101$ MPa at $T = 1273$ K; and the interface debonding length decreases from $l_d/r_f = 4.8$ at $\sigma_{mc} = 121$ MPa to $l_d/r_f = 4.1$ at $\sigma_{mc} = 101$ MPa.

- When the interface debonding energy is $\Gamma_i = 0.3$ J/m^2, the FMCS decreases from $\sigma_{mc} = 125$ MPa at $T = 873$ K to $\sigma_{mc} = 103$ MPa at $T = 1273$ K; and the interface debonding length decreases from $l_d/r_f = 4.5$ at $\sigma_{mc} = 125$ MPa to $l_d/r_f = 3.9$ at $\sigma_{mc} = 103$ MPa.

Effect of the Matrix Fracture Energy on Temperature-Dependent FMCS
The FMCS and the fiber/matrix interface debonding length versus the temperature curves for different matrix fracture energies (i.e., $\Gamma_m = 15, 20,$ and 25 J/m^2) are shown in Figure 2.8. When the temperature increases from $T = 873$ to 1273 K, the FMCS and interface debonding length decrease with the increase in temperature. At the same

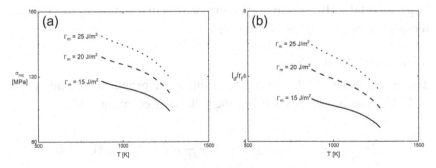

FIGURE 2.8 Effect of the matrix fracture energy on (a) the FMCS versus temperature curves; (b) the fiber/matrix interface debonding length versus temperature curves of SiC/SiC composites.

temperature, the FMCS and the interface debonding length increase with the matrix fracture energy. When the matrix fracture energy increases, the energy needed for the matrix cracking increases, leading to the increase in the FMCS and the interface debonding length.

- When the matrix fracture energy is $\Gamma_m = 15$ J/m^2, the FMCS decreases from $\sigma_{mc} = 117$ MPa at $T = 873$ K to $\sigma_{mc} = 99$ MPa at $T = 1273$ K; and the interface debonding length decreases from $l_d/r_f = 5.3$ at $\sigma_{mc} = 117$ MPa to $l_d/r_f = 4.4$ at $\sigma_{mc} = 99$ MPa.

- When the matrix fracture energy is $\Gamma_m = 20$ J/m^2, the FMCS decreases from $\sigma_{mc} = 132$ MPa at $T = 873$ K to $\sigma_{mc} = 110$ MPa at $T = 1273$ K; and the interface debonding length decreases from $l_d/r_f = 6.2$ at $\sigma_{mc} = 132$ MPa to $l_d/r_f = 5$ at $\sigma_{mc} = 110$ MPa.

- When the matrix fracture energy is $\Gamma_m = 25$ J/m^2, the FMCS decreases from $\sigma_{mc} = 145$ MPa at $T = 873$ K to $\sigma_{mc} = 119$ MPa at $T = 1273$ K; and the interface debonding length decreases from $l_d/r_f = 6.9$ at $\sigma_{mc} = 145$ MPa to $l_d/r_f = 5.5$ at $\sigma_{mc} = 119$ MPa.

TIME-DEPENDENT FMCS IN SiC/SiC COMPOSITES

In this section, the synergistic effects of temperature and time on the FMCS of SiC/SiC composites are investigated. The temperature-dependent constituent properties and time-dependent interface oxidation are considered to determine the interface debonded length and FMCS. Effects of fiber volume, ISS, interface debonded energy, and matrix fracture energy on the temperature/time-dependent FMCS and interface debonded length of SiC/SiC composites are discussed. Experimental FMCS of SiC/SiC composites corresponding to different temperatures and times is predicted.

Micromechanical Model

The energy balance relationship to evaluate the proportional limit stress of CMCs can be described using the following equation [18]:

$$\frac{1}{2}\int_{-\infty}^{\infty}\left\{\frac{V_f}{E_f(T)}[\sigma_{fu}(T) - \sigma_{fd}(T)]^2 + \frac{V_m}{E_m(T)}[\sigma_{mu}(T) - \sigma_{md}(T)]^2\right\}dx$$

$$+ \frac{1}{2\pi R^2 G_m(T)}\int_{-l_d(T)}^{l_d(T)}\int_{r_f}^{R}\left[\frac{r_f\tau_i(x,T)}{r}\right]2\pi r dr dx \qquad (2.17)$$

$$= V_m\Gamma_m(T) + \frac{4V_f l_d(T)}{r_f}\Gamma_i(T)$$

where V_f and V_m denote the fiber and matrix volume fraction, respectively; $E_f(T)$ and $E_m(T)$ denote the temperature-dependent fiber and matrix elastic modulus, respectively; $\sigma_{fu}(T)$ and $\sigma_{mu}(T)$ denote the fiber and matrix axial stress distribution in the matrix cracking upstream region, respectively; $\sigma_{fd}(T)$ and $\sigma_{md}(T)$ denote the fiber and matrix axial stress distribution in the matrix cracking downstream region, respectively; and $\Gamma_m(T)$ and $\Gamma_i(T)$ denote the temperature-dependent matrix fracture energy and interface debonded energy, respectively.

$$\sigma_{fu}(T) = \frac{E_f(T)}{E_c(T)}\sigma \qquad (2.18)$$

$$\sigma_{mu}(T) = \frac{E_m(T)}{E_c(T)}\sigma \qquad (2.19)$$

$$\sigma_{fd}(x, T) = \begin{cases} \frac{\sigma}{V_f} - \frac{2\tau_f(T)}{r_f}x, & x \in [0, \zeta(T)] \\ \frac{\sigma}{V_f} - \frac{2\tau_f(T)}{r_f}\zeta(T) - \frac{2\tau_i(T)}{r_f}[x - \zeta(T)], & x \in [\zeta(T), l_d(T)] \\ \frac{E_f(T)}{E_c(T)}\sigma, & x \in \left[l_d(T), \frac{l_c(T)}{2}\right] \end{cases}$$

$$(2.20)$$

$$\sigma_{md}(x, T) = \begin{cases} 2\frac{V_f}{V_m}\frac{\tau_f(T)}{r_f}x, & x \in [0, \zeta(T)] \\ 2\frac{V_f}{V_m}\frac{\tau_f(T)}{r_f}\zeta(T) + 2\frac{V_f}{V_m}\frac{\tau_i(T)}{r_f}[x - \zeta(T)], & x \in [\zeta(T), l_d(T)] \\ \frac{E_m(T)}{E_c(T)}\sigma, & x \in \left[l_d(T), \frac{l_c(T)}{2}\right] \end{cases}$$

$$(2.21)$$

$$l_d(T) = \left[1 - \frac{\tau_f(T)}{\tau_i(T)}\right]\zeta(T) + \frac{r_f}{2}\frac{V_m E_m(T)\sigma}{V_f E_c(T)\tau_i(T)}$$
$$- \sqrt{\frac{r_f V_m E_m(T)E_f(T)}{E_c(T)\tau_i^2(T)}\gamma_d(T)} \qquad (2.22)$$

where

$$\zeta = \varphi_1\left[1 - \exp\left(-\frac{\varphi_2 t}{b}\right)\right] \qquad (2.23)$$

$$\tau_i(T) = \tau_0 + \mu \frac{|\alpha_{rf}(T) - \alpha_{rm}(T)|(T_m - T)}{A} \tag{2.24}$$

$$\tau_f(T) = \tau_s + \mu \frac{|\alpha_{rf}(T) - \alpha_{rm}(T)|(T_m - T)}{A} \tag{2.25}$$

Substituting the upstream and downstream temperature-dependent fiber and matrix axial stresses of Eqs. (2.18), (2.19), (2.20), and (2.21) and the temperature-dependent fiber/matrix interface debonded length of Eq. (2.22) into Eq. (2.17), the energy balance equation leads to the following equation:

$$\alpha \sigma^2 + \beta \sigma + \delta = 0 \tag{2.26}$$

where

$$\alpha = \frac{V_m E_m(T) l_d(T)}{V_f E_f(T) E_c(T)} \tag{2.27}$$

$$\beta = -\frac{2\tau_i(T)}{r_f E_f(T)} \left\{ [l_d(T) - \zeta(T)]^2 + \frac{\tau_f(T)}{\tau_i(T)} \zeta(T)[2l_d(T) - \zeta(T)] \right\} \tag{2.28}$$

$$
\begin{aligned}
\delta = {} & \frac{4}{3} \frac{V_f E_c(T)}{V_m E_f(T) E_m(T)} \left[\frac{\tau_i(T)}{r_f} \right]^2 \left\{ [l_d(T) - \zeta(T)]^3 + \left[\frac{\tau_f(T)}{\tau_i(T)} \right]^2 \zeta^3(T) \right\} \\
& + \frac{4 V_f E_c(T)}{V_m E_f(T) E_m(T)} \frac{\tau_f(T)\tau_i(T)}{r_f^2} \zeta(T)[l_d(T) - \zeta(T)] \\
& \times \left\{ l_d(T) - \left(1 - \frac{\tau_f(T)}{\tau_i(T)}\right) \zeta(T) \right\} - V_m \Gamma_m(T) - \frac{4 V_f l_d(T)}{r_f} \Gamma_i(T)
\end{aligned}
\tag{2.29}
$$

Experimental Comparisons

Guo and Kagawa [29] investigated the tensile behavior of 2D SiC/SiC composites with PyC and BN interphase at elevated temperature. The experimental and predicted FMCS versus temperature curves are shown in Figure 2.9.

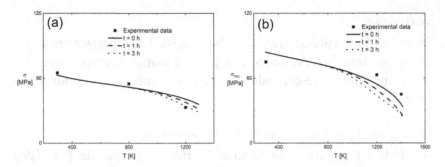

FIGURE 2.9 The experimental and predicted FMCS versus the temperature curves of (a) SiC/SiC composites with the PyC interphase and (b) SiC/SiC composites with the BN interphase.

- For 2D SiC/SiC composites with the PyC interphase without oxidation, the experimental FMCS decreases from $\sigma_{mc} = 65$ MPa at $T = 298$ K to $\sigma_{mc} = 33$ MPa at $T = 1200$ K, the predicted FMCS decreases from $\sigma_{mc} = 63$ MPa at $T = 298$ K to $\sigma_{mc} = 35$ MPa at $T = 1200$ K, and the interface debonded length decreases from $l_d/r_f = 12.8$ to $l_d/r_f = 7.3$; when the oxidation time is $t = 1$ h, the FMCS decreases from $\sigma_{mc} = 63$ MPa at $T = 298$ K to $\sigma_{mc} = 31$ MPa at $T = 1200$ K, and the interface debonded length decreases from $l_d/r_f = 12.8$ to $l_d/r_f = 7.7$; when the oxidation time is $t = 3$ h, the FMCS decreases from $\sigma_{mc} = 63$ MPa at $T = 298$ K to $\sigma_{mc} = 28$ MPa at $T = 1200$ K, and the interface debonded length decreases from $l_d/r_f = 12.8$ to $l_d/r_f = 9.8$.

- For 2D SiC/SiC composites with the BN interphase without oxidation, the experimental FMCS decreases from $\sigma_{mc} = 75$ MPa at $T = 298$ K to $\sigma_{mc} = 45$ MPa at $T = 1400$ K, the predicted FMCS decreases from $\sigma_{mc} = 84$ MPa at $T = 298$ K to $\sigma_{mc} = 33$ MPa at $T = 1400$ K, and the interface debonded length decreases from $l_d/r_f = 7.3$ to $l_d/r_f = 2.9$; when the oxidation time is $t = 1$ h, the FMCS decreases from $\sigma_{mc} = 84$ MPa at $T = 298$ K to $\sigma_{mc} = 25$ MPa at $T = 1400$ K, and the interface debonded length decreases from $l_d/r_f = 7.3$ to $l_d/r_f = 3.8$; when the oxidation time is $t = 3$ h, the FMCS decreases from $\sigma_{mc} = 84$ MPa at $T = 298$ K to $\sigma_{mc} = 23$ MPa at $T = 1400$ K, and the interface debonded length decreases from $l_d/r_f = 7.3$ to $l_d/r_f = 7.1$.

Discussions

Effects of fiber volume fraction, fiber/matrix ISS, interface frictional coefficient, interface debonded energy, and matrix fracture energy on the temperature/time-dependent FMCS and interface debonding are discussed.

Effect of the Fiber Volume on Time-Dependent FMCS

The FMCS (σ_{mc}) and fiber/matrix interface debonded length (l_d/r_f) versus temperature curves of SiC/SiC composites corresponding to different fiber volume fractions of $V_f = 25\%$ and 30% and oxidation times of $t = 0$, 1, and 3 h are shown in Figure 2.10.

- When the fiber volume fraction is $V_f = 25\%$ and without oxidation, the FMCS decreases from $\sigma_{mc} = 100$ MPa at $T = 973$ K to $\sigma_{mc} = 87$ MPa at $T = 1273$ K, and the interface debonded length decreases from $l_d/r_f = 5.8$ at $T = 973$ K to $l_d/r_f = 5$ at $T = 1273$ K; when the

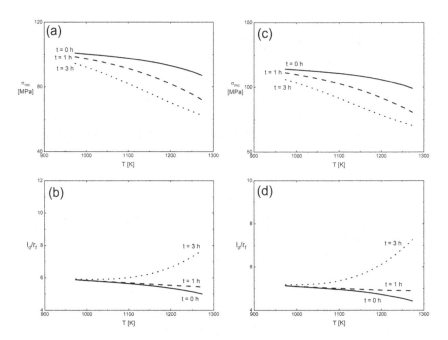

FIGURE 2.10 (a) The FMCS versus the temperature curves when $V_f = 25\%$ and $t = 0$, 1, and 3 h; (b) the interface debonded length versus the temperature curves when $V_f = 25\%$ and $t = 0$, 1, and 3 h; (c) the FMCS versus the temperature curves when $V_f = 30\%$ and $t = 0$, 1, and 3 h; and (d) the interface debonded length versus the temperature curves when $V_f = 30\%$ and $t = 0$, 1, and 3 h.

oxidation time is $t = 1$ h, the FMCS decreases from $\sigma_{mc} = 98$ MPa at $T = 973$ K to $\sigma_{mc} = 72$ MPa at $T = 1273$ K, and the interface debonded length decreases from $l_d/r_f = 5.8$ at $T = 973$ K to $l_d/r_f = 5.4$ at $T = 1273$ K; and when the oxidation time is $t = 3$ h, the FMCS decreases from $\sigma_{mc} = 94$ MPa at $T = 973$ K to $\sigma_{mc} = 62$ MPa at $T = 1273$ K, and the interface debonded length increases from $l_d/r_f = 5.9$ at $T = 973$ K to $l_d/r_f = 7.6$ at $T = 1273$ K.

- When the fiber volume fraction is $V_f = 30\%$ and without oxidation, the FMCS increases from $\sigma_{mc} = 114$ MPa at $T = 973$ K to $\sigma_{mc} = 99$ MPa at $T = 1273$ K, and the interface debonded length decreases from $l_d/r_f = 5.1$ at $T = 973$ K to $l_d/r_f = 4.4$ at $T = 1273$ K; when the oxidation time is $t = 1$ h, the FMCS decreases from $\sigma_{mc} = 111$ MPa at $T = 973$ K to $\sigma_{mc} = 80$ MPa at $T = 1273$ K, and the interface debonded length decreases from $l_d/r_f = 5.1$ at $T = 973$ K to $l_d/r_f = 4.9$ at $T = 1273$ K; and when the oxidation time is $t = 3$ h, the FMCS decreases from $\sigma_{mc} = 105$ MPa at $T = 973$ K to $\sigma_{mc} = 70$ MPa at $T = 1273$ K, and the interface debonded length increases from $l_d/r_f = 5.1$ at $T = 973$ K to $l_d/r_f = 7.2$ at $T = 1273$ K.

Effect of the ISS on Time-Dependent FMCS

The FMCS (σ_{mc}) and fiber/matrix interface debonded length (l_d/r_f) versus temperature curves of SiC/SiC composites corresponding to different ISS of $\tau_0 = 15$ and 20 MPa and oxidation times of $t = 0$, 1, and 3 h are shown in Figure 2.11.

- When the ISS is $\tau_0 = 15$ MPa and without oxidation, the FMCS decreases from $\sigma_{mc} = 88.3$ MPa at $T = 973$ K to $\sigma_{mc} = 80.9$ MPa at $T = 1273$ K, and the fiber/matrix interface debonded length decreases from $l_d/r_f = 6.6$ at $T = 973$ K to $l_d/r_f = 5.9$ at $T = 1273$ K; when the oxidation time is $t = 1$ h, the FMCS decreases from $\sigma_{mc} = 87.5$ MPa at $T = 973$ K to $\sigma_{mc} = 74.6$ MPa at $T = 1273$ K, and the interface debonded length decreases from $l_d/r_f = 6.6$ at $T = 973$ K to $l_d/r_f = 6.1$ at $T = 1273$ K; and when the oxidation time is $t = 2$ h, the FMCS decreases from $\sigma_{mc} = 86.1$ MPa at $T = 973$ K to $\sigma_{mc} = 70.2$ MPa at $T = 1273$ K, and the interface debonded length increases from $l_d/r_f = 6.6$ at $T = 973$ K to $l_d/r_f = 7.2$ at $T = 1273$ K.

- When the ISS is $\tau_0 = 20$ MPa and without oxidation, the FMCS decreases from $\sigma_{mc} = 102$ MPa at $T = 973$ K to $\sigma_{mc} = 90$ MPa at $T =$

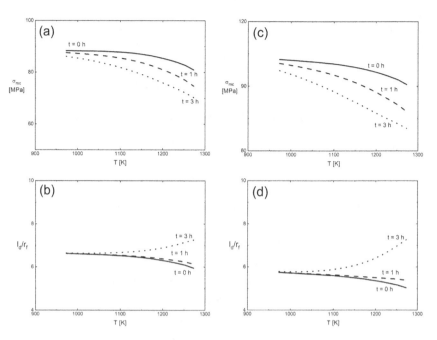

FIGURE 2.11 (a) The FMCS versus the temperature curves when $\tau_0 = 15$ MPa and $t = 0$, 1, and 3 h; (b) the interface debonded length versus the temperature curves when $\tau_0 = 15$ MPa and $t = 0$, 1, and 3 h; (c) the FMCS versus the temperature curves when $\tau_0 = 20$ MPa and $t = 0$, 1, and 3 h; and (d) the interface debonded length versus the temperature curves when $\tau_0 = 20$ MPa and $t = 0$, 1, and 3 h.

1273 K, and the fiber/matrix interface debonded length decreases from $l_d/r_f = 5.7$ at $T = 973$ K to $l_d/r_f = 5.0$ at $T = 1273$ K; when the oxidation time is $t = 1$ h, the FMCS decreases from $\sigma_{mc} = 100.5$ MPa at $T = 973$ K to $\sigma_{mc} = 78$ MPa at $T = 1273$ K, and the fiber/matrix interface debonded length decreases from $l_d/r_f = 5.7$ at $T = 973$ K to $l_d/r_f = 5.3$ at $T = 1273$ K; and when the oxidation time is $t = 2$ h, the FMCS decreases from $\sigma_{mc} = 97.2$ MPa at $T = 973$ K to $\sigma_{mc} = 70.4$ MPa at $T = 1273$ K, and the fiber/matrix interface debonded length increases from $l_d/r_f = 5.7$ at $T = 973$ K to $l_d/r_f = 7.2$ at $T = 1273$ K.

The FMCS (σ_{mc}) and fiber/matrix interface debonded length (l_d/r_f) versus temperature curves of SiC/SiC composites corresponding to different ISS of $\tau_s = 10$ and 15 MPa and oxidation times of $t = 0$, 1, and 3 h are shown in Figure 2.12.

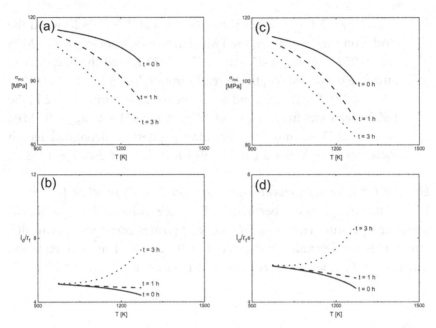

FIGURE 2.12 (a) The FMCS versus the temperature curves when $\tau_s = 10$ MPa and $t = 0$, 1, and 3 h; (b) the interface debonded length versus the temperature curves when $\tau_s = 10$ MPa and $t = 0$, 1, and 3 h; (c) the FMCS versus the temperature curves when $\tau_s = 15$ MPa and $t = 0$, 1, and 3 h; and (d) the interface debonded length versus the temperature curves when $\tau_0 = 15$ MPa and $t = 0$, 1, and 3 h.

- When the ISS is $\tau_s = 10$ MPa and without oxidation, the FMCS decreases from $\sigma_{mc} = 114$ MPa at $T = 973$ K to $\sigma_{mc} = 99$ MPa at $T = 1273$ K, and the fiber/matrix interface debonded length decreases from $l_d/r_f = 5.1$ at $T = 973$ K to $l_d/r_f = 4.4$ at $T = 1273$ K; when the oxidation time is $t = 1$ h, the FMCS decreases from $\sigma_{mc} = 111$ MPa at $T = 973$ K to $\sigma_{mc} = 80$ MPa at $T = 1273$ K, and the fiber/matrix interface debonded length decreases from $l_d/r_f = 5.1$ at $T = 973$ K to $l_d/r_f = 4.9$ at $T = 1273$ K; and when the oxidation time is $t = 2$ h, the FMCS decreases from $\sigma_{mc} = 106$ MPa at $T = 973$ K to $\sigma_{mc} = 70$ MPa at $T = 1273$ K, and the fiber/matrix interface debonded length increases from $l_d/r_f = 5.1$ at $T = 973$ K to $l_d/r_f = 7.2$ at $T = 1273$ K.

- When the ISS is $\tau_s = 15$ MPa and without oxidation, the FMCS decreases from $\sigma_{mc} = 114$ MPa at $T = 973$ K to $\sigma_{mc} = 99$ MPa at $T = 1273$ K, and the fiber/matrix interface debonded length decreases from l_d/r_f = 5.1 at T = 973 K to l_d/r_f = 4.4 at T = 1273 K; when the oxidation time is t = 1 h, the FMCS decreases from σ_{mc} = 112 MPa at T = 973 K to σ_{mc} = 87 MPa at T = 1273 K, and the fiber/matrix interface debonded length decreases from l_d/r_f = 5.1 at T = 973 K to l_d/r_f = 4.7 at T = 1273 K; and when the oxidation time is t = 2 h, the FMCS decreases from σ_{mc} = 108 MPa at T = 973 K to σ_{mc} = 82 MPa at T = 1273 K, and the fiber/matrix interface debonded length increases from l_d/r_f = 5.1 at T = 973 K to l_d/r_f = 6.3 at T = 1273 K.

Effect of the Interface Debonding Energy on Time-Dependent FMCS
The FMCS (σ_{mc}) and fiber/matrix interface debonded length (l_d/r_f) versus temperature curves of SiC/SiC composites corresponding to different interface debonded energies of Γ_i = 0.1 and 0.3 J/m^2 and oxidation times of t = 0, 1, and 2 h are shown in Figure 2.13.

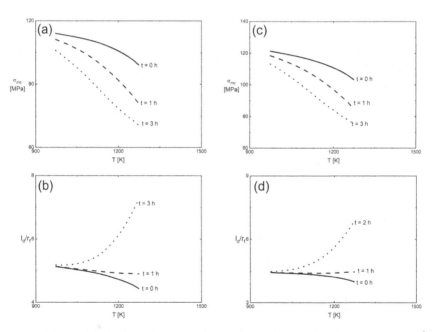

FIGURE 2.13 (a) The FMCS versus the temperature curves when Γ_i = 0.1 J/m^2 and t = 0, 1, and 3 h; (b) the interface debonded length versus the temperature curves when Γ_i = 0.1 J/m^2 and t = 0, 1, and 3 h; (c) the FMCS versus the temperature curves when Γ_i = 0.3 J/m^2 and t = 0, 1, and 3 h; and (d) the interface debonded length versus the temperature curves when Γ_i = 0.3 J/m^2 and t = 0, 1, and 3 h.

- When the interface debonded energy is $\Gamma_i = 0.1$ J/m^2 and without oxidation, the FMCS decreases from $\sigma_{mc} = 114$ MPa at $T = 973$ K to $\sigma_{mc} = 99$ MPa at $T = 1273$ K, and the fiber/matrix interface debonded length decreases from $l_d/r_f = 5.1$ at $T = 973$ K to $l_d/r_f = 4.4$ at $T = 1273$ K; when the oxidation time is $t = 1$ h, the FMCS decreases from $\sigma_{mc} = 111$ MPa at $T = 973$ K to $\sigma_{mc} = 80$ MPa at $T = 1273$ K, and the fiber/matrix interface debonded length decreases from $l_d/r_f = 5.1$ at $T = 973$ K to $l_d/r_f = 4.9$ at $T = 1273$ K; and when the oxidation time is $t = 2$ h, the FMCS decreases from $\sigma_{mc} = 105$ MPa at $T = 973$ K to $\sigma_{mc} = 70$ MPa at $T = 1273$ K, and the fiber/matrix interface debonded length increases from $l_d/r_f = 5.1$ at $T = 973$ K to $l_d/r_f = 7.2$ at $T = 1273$ K.

- When the interface debonded energy is $\Gamma_i = 0.3$ J/m^2 and without oxidation, the FMCS decreases from $\sigma_{mc} = 121$ MPa at $T = 973$ K to $\sigma_{mc} = 103$ MPa at $T = 1273$ K, and the fiber/matrix interface debonded length decreases from $l_d/r_f = 4.4$ at $T = 973$ K to $l_d/r_f = 3.9$ at $T = 1273$ K; when the oxidation time is $t = 1$ h, the FMCS decreases from $\sigma_{mc} = 118$ MPa at $T = 973$ K to $\sigma_{mc} = 84$ MPa at $T = 1273$ K, and the fiber/matrix interface debonded length increases from $l_d/r_f = 4.3$ at $T = 973$ K to $l_d/r_f = 4.4$ at $T = 1273$ K; and when the oxidation time is $t = 2$ h, the FMCS decreases from $\sigma_{mc} = 113$ MPa at $T = 973$ K to $\sigma_{mc} = 75$ MPa at $T = 1273$ K, and the fiber/matrix interface debonded length increases from $l_d/r_f = 4.4$ at $T = 973$ K to $l_d/r_f = 6.8$ at $T = 1273$ K.

Effect of the Matrix Fracture Energy on Time-Dependent FMCS

The FMCS (σ_{mc}) and fiber/matrix interface debonded length (l_d/r_f) versus temperature curves of C/SiC composites corresponding to different matrix fracture energies of $\Gamma_m = 15$ and 20 J/m^2 and oxidation times of $t = 0$, 1, and 2 h are shown in Figure 2.14.

- When the matrix fracture energy is $\Gamma_m = 15$ J/m^2 and without oxidation, the FMCS decreases from $\sigma_{mc} = 114$ MPa at $T = 973$ K to $\sigma_{mc} = 99$ MPa at $T = 1273$ K, and the fiber/matrix interface debonded length decreases from $l_d/r_f = 5.1$ at $T = 973$ K to $l_d/r_f = 4.4$ at $T = 1273$ K; when the oxidation time is $t = 1$ h, the FMCS decreases from $\sigma_{mc} = 111$ MPa at $T = 973$ K to $\sigma_{mc} = 80$ MPa at $T = 1273$ K, and the fiber/matrix interface debonded length

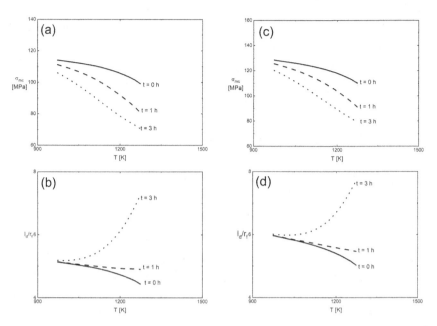

FIGURE 2.14 (a) The FMCS versus the temperature curves when $\Gamma_m = 15$ J/m^2 and $t = 0$, 1, and 3 h; (b) the interface debonded length versus the temperature curves when $\Gamma_m = 15$ J/m^2 and $t = 0$, 1, and 3 h; (c) the FMCS versus the temperature curves when $\Gamma_m = 20$ J/m^2 and $t = 0$, 1, and 3 h; and (d) the interface debonded length versus the temperature curves when $\Gamma_m = 20$ J/m^2 and $t = 0$, 1, and 3 h.

decreases from $l_d/r_f = 5.1$ at $T = 973$ K to $l_d/r_f = 4.9$ at $T = 1273$ K; and when the oxidation time is $t = 2$ h, the FMCS decreases from $\sigma_{mc} = 106$ MPa at $T = 973$ K to $\sigma_{mc} = 70$ MPa at $T = 1273$ K, and the fiber/matrix interface debonded length increases from $l_d/r_f = 5.1$ at $T = 973$ K to $l_d/r_f = 7.2$ at $T = 1273$ K.

• When the matrix fracture energy is $\Gamma_m = 20$ J/m^2 and without oxidation, the FMCS decreases from $\sigma_{mc} = 128$ MPa at $T = 973$ K to $\sigma_{mc} = 110$ MPa at $T = 1273$ K, and the fiber/matrix interface debonded length decreases from $l_d/r_f = 5.9$ at $T = 973$ K to $l_d/r_f = 5.0$ at $T = 1273$ K; when the oxidation time is $t = 1$ h, the FMCS decreases from $\sigma_{mc} = 125$ MPa at $T = 973$ K to $\sigma_{mc} = 91$ MPa at $T = 1273$ K, and the fiber/matrix interface debonded length decreases from $l_d/r_f = 5.9$ at $T = 973$ K to $l_d/r_f = 5.4$ at $T = 1273$ K; and when the oxidation time is $t = 2$ h, the FMCS decreases from $\sigma_{mc} = 120$ MPa at $T = 973$ K to $\sigma_{mc} = 79$ MPa at $T = 1273$ K, and the fiber/matrix interface debonded length increases from $l_d/r_f = 6.0$ at $T = 973$ K to $l_d/r_f = 7.7$ at $T = 1273$ K.

SUMMARY AND CONCLUSIONS

In this chapter, the effects of temperature and time on the FMCS of SiC/SiC composites were investigated. The temperature-dependent constituent properties and time-dependent interface oxidation were considered to determine the FMCS. Effects of interface oxidation and constituent properties on the temperature/time-dependent FMCS and interface debonded length of SiC/SiC composite were discussed. Experimental FMCS of SiC/SiC composites corresponding to different temperatures and times was predicted.

1. When the fiber volume fraction, ISS, and interface debonded energy increase, the FMCS at the same temperature increases, and the fiber/matrix interface debonded length decreases.

2. When the matrix fracture energy increases, the FMCS and the interface debonded length increase at the same temperature.

3. When the oxidation time increases, the FMCS decreases, and the interface debonded length increases.

REFERENCES

1. Li LB. *Damage, fracture and fatigue of ceramic-matrix composites.* Springer Nature Singapore Pte Ltd., 2018.
2. Naslain R. Design, preparation and properties of non-oxide CMCs for application in engines and nuclear reactors: An overview. *Compos. Sci. Technol.* 2004; 64:155–170.
3. Kotani M, Konaka K, Ogihara S. The effect on the tensile properties of PIP-processed SiC/SiC composite of a chemical vapor-infiltrated SiC layer overlaid on the pyrocarbon interface layer. *Compos. A* 2016; 87:123–130.
4. Kabel J, Yang Y, Balooch M, Howard C, Koyanagi T, Terrani KA, Katoh Y, Hosemann P. Micro-mechanical evaluation of SiC-SiC composite interphase properties and debond mechanisms. *Compos. B* 2017; 131:173–183.
5. Hasegawa A, Kohyama A, Jones RH, Snead LL, Riccardi B, Fenici P. Critical issues and current status of SiC/SiC composites for fusion. *J. Nucl. Mater.* 2000; 283-287:128–137.

6. Ivekovic A, Novak S, Drazic G, Blagoeva D, Vicente S. Current status and prospects of SiC$_f$/SiC for fusion structural applications. *J. Eur. Ceram. Soc.* 2013; 33:1577–1589.

7. Udayakumar A, Stalin M, Abhayalakshmi MB, Hariharan R, Balasubramanian M. Effect of thermal cycling of SiC$_f$/SiC composites on their mechanical properties. *J. Nucl. Mater.* 2013; 442:S384–S389.

8. Katoh Y, Ozawa K, Shih C, Nozawa T, Shinavski RJ, Hasegawa A, Snead LL. Continuous SiC fiber, CVI SiC matrix composites for nuclear applications: Properties and irradiation effects. *J. Nucl. Mater.* 2014; 448:448–476.

9. Singh G, Terrani K, Katoh Y. Thermo-mechanical assessment of full SiC/SiC composite cladding for LWR applications with sensitivity analysis. *J. Nucl. Mater.* 2018; 499:126–143.

10. Kimmel J, Miriyala N, Price J, More K, Tortorelli P, Eaton H, Linsey G, Sun E. Evaluation of CFCC liners with EBC after field testing in a gas turbine. *J. Eur. Ceram. Soc.* 2002; 22:2769–2775.

11. Filsinger D, Munz S, Schulz A, Wittig S, Andrees G. Experimental assessment of fiber-reinforced ceramics for combustor walls. *J. Eng. Gas Turbines Power* 2001; 123:271–276.

12. Li LB. Micromechanical modeling for tensile behavior of carbon fiber-reinforced ceramic-matrix composites. *Appl. Compos. Mater.* 2015; 22:773–790.

13. Morscher GN, Singh M, Kiser JD, Freedman M, Bhatt R. Modeling stress-dependent matrix cracking and stress–strain behavior in 2D woven SiC fiber reinforced CVI SiC composites. *Compos. Sci. Technol.* 2007; 67:1009–1017.

14. Li L, Song Y, Sun Y. Modeling tensile behavior of cross-ply C/SiC ceramic-matrix composites. *Mech. Compos. Mater.* 2015; 51:358–376.

15. Li L, Song Y, Sun Y. Modeling tensile behavior of unidirectional C/SiC ceramic matrix composites. *Mech. Compos. Mater.* 2014; 49:659–672.

16. Morscher GN, Gordon NA. Acoustic emission and electrical resistance in SiC-based laminate ceramic composites tested under tensile loading. *J. Eur. Ceram. Soc.* 2017; 37:3861–3872.

17. Aveston J, Cooper GA, Kelly A. Single and multiple fracture. In: The Properties of Fiber Composites, Conference on Proceedings. National Physical Laboratory. Guildford: IPC Science and Technology Press, 1971; 15–26.

18. Budiansky B, Hutchinson JW, Evans AG. Matrix fracture in fiber-reinforced ceramics. *J. Mech. Phys. Solids* 1986; 34:167–189.

19. Sutcu M, Hillig WB. The effects of fiber-matrix debond energy on the matrix cracking strength and debond shear strength. *Acta Metall. Mater.* 1990; 38:2653–2662.

20. Chiang Y, Wang ASD, Chou TW. On matrix cracking in fiber reinforced ceramics. *J. Mech. Phys. Solids* 1993; 41:1137–1154.

21. Chiang Y. On fiber debonding and matrix cracking in fiber-reinforced ceramics. *Compos. Sci. Technol.* 2001; 61:1743–1756.

22. Chiang Y. On a matrix cracking model using Coulomb's friction law. *Eng. Fract. Mech.* 2007; 74:1602–1616.
23. Marshall DB, Cox BN, Evans AG. The mechanics of matrix cracking in brittle-matrix fiber composites. *Acta Metall.* 1985; 33:2013–2021.
24. Marshall DB, Cox BN. Tensile fracture of brittle matrix composites: Influence of fiber strength. *Acta Metall.* 1987; 35:2607–2619.
25. Chiang Y. Tensile failure in fiber reinforced ceramic matrix composites. *J. Mater. Sci.* 2000; 35:5449–5455.
26. Pavia F, Letertre A, Curtin WA. Prediction of first matrix cracking in micro/nanohybrid brittle matrix composites. *Compos. Sci. Technol.* 2010; 70:916–921.
27. Li LB. Synergistic effects of fiber/matrix interface wear and fibers fracture on matrix multiple cracking in fiber-reinforced ceramic-matrix composites. *Compos. Interfaces* 2019; 26:193–219.
28. Singh RN. Influence of interfacial shear stress on first-matrix cracking stress in ceramic-matrix composites. *J. Am. Ceram. Soc.* 1990; 73:2930–2937.
29. Guo S, Kagawa Y. Tensile fracture behavior of continuous SiC fiber-reinforced SiC matrix composites at elevated temperatures and correlation to in situ constituent properties. *J. Eur. Ceram. Soc.* 2002; 22:2349–2356.
30. Fantozzi G, Reynaud P, Rouby D. Thermomechanical behavior of long fibers ceramic-ceramic composites. *Silic. Ind.* 2001; 66:109–119.
31. Snead LL, Nozawa T, Katoh Y, Byun TS, Kondo S, Petti DA. Handbook of SiC properties for fuel performance modeling. *J. Nucl. Mater.* 2007; 371:329–377.
32. Wang RZ, Li WG, Li DY, Fang DN. A new temperature dependent fracture strength model for the ZrB_2-SiC composites. *J. Eur. Ceram. Soc.* 2015; 35:2957–2962.

CHAPTER 3

High-Temperature Multiple Matrix Cracking Behavior in Ceramic-Matrix Composites

INTRODUCTION

At high temperatures, matrix cracking is an important failure mode in ceramic-matrix composites (CMCs) [1–7]. The generation and propagation of matrix cracking consume the energy inside CMCs, which slows down or prevents further matrix cracking propagation and achieves toughness behavior [8–12]. Nonlinearity of stress–strain curves of CMCs is mainly caused by matrix cracking and propagation [13–23]. Temperature dependence of composite's constituent properties, i.e., interface shear stress, Young's modulus of matrix and fiber, matrix fracture energy, and interface debonding energy, affects the evolution of matrix multiple cracking [24]. Matrix cracks can form paths for the ingress of the environment, oxidizing the fibers, which leads to premature failure [25–29]. The density and openings of these cracks depend on the fiber architecture, fiber/matrix interface bonding intensity, applied load, and environments [30–33].

In this chapter, the effects of temperature and time on the matrix multiple cracking in CMCs were investigated. A critical matrix strain energy (CMSE) criterion was adopted to determine the matrix cracking density as a function of applied stress, temperature, and duration at high

DOI: 10.1201/9781032638508-3

temperatures. Experimental matrix cracking densities of SiC/SiC composites corresponding to different temperatures and times were predicted. Effects of fiber volume fraction, interface properties, and matrix property on the temperature/time-dependent matrix multiple cracking of SiC/SiC composites were also discussed.

TEMPERATURE-DEPENDENT MULTIPLE MATRIX CRACKING EVOLUTION IN SiC/SiC COMPOSITES

In this section, the temperature-dependent multiple matrix cracking evolution in SiC/SiC composites was investigated using the CMSE criterion. The BHE shear-lag model was used to describe the micro-stress field of the damaged composite. A fracture mechanics approach and the CMSE criterion were adopted to determine the interface debonding length and matrix cracking density. Experimental matrix multiple cracking evolution of SiC/SiC composites at high temperatures was predicted. The effect of composite's constituent properties on matrix multiple cracking evolution of SiC/SiC composites was also discussed at different testing temperatures.

Micromechanical Model

The CMSE criterion was adopted to determine the matrix multiple cracking evolution in CMCs. The temperature-dependent matrix strain energy $U_m(T)$ for the partial interface debonding state is

$$
\begin{aligned}
U_m(T) = \frac{A_m}{E_m(T)} &\left\{ \frac{4}{3}\left[\frac{V_f \tau_i(T)}{V_m r_f} l_d(T) \right]^2 l_d(T) + \sigma_{mo}^2(T)\left[\frac{l_c(T)}{2} - l_d(T) \right] \right. \\
&+ 2\frac{r_f \sigma_{mo}(T)}{\rho}\left[\frac{2V_f \tau_i(T) l_d(T)}{V_m r_f} - \sigma_{mo}(T) \right] \\
&\left[1 - \exp\left(-\rho\frac{l_c(T)/2 - l_d(T)}{r_f} \right) \right] \\
&\left. + \frac{r_f}{2\rho}\left[\frac{2V_f \tau_i(T) l_d(T)}{V_m r_f} - \sigma_{mo}(T) \right]^2 \left[1 - \exp\left(-2\rho\frac{l_c(T)/2 - l_d(T)}{r_f} \right) \right] \right\}
\end{aligned}
$$

(3.1)

where A_m is the cross-section of the matrix; V_f and V_m denote the fiber and matrix volume fraction, respectively; r_f is the fiber radius; $E_m(T)$ is the temperature-dependent matrix modulus; $l_d(T)$ and $l_c(T)$ denote the temperature-dependent interface debonding length and matrix crack spacing, respectively; ρ is the shear-lag model parameter; and $\sigma_{mo}(T)$ denotes the temperature-dependent matrix axial stress.

For the damage state of complete interface debonding, the temperature-dependent matrix strain energy $U_m(T)$ is

$$U_m(T) = \frac{A_m V_f^2 l_c^3(T) \tau_i^2(T)}{6r_f^2 V_m^2 E_m(T)} \tag{3.2}$$

The temperature-dependent CMSE $U_m^{cr}(T)$ is

$$U_m^{cr}(T) = \frac{1}{2} k A_m l_0 \frac{\sigma_{mocr}^2(T)}{E_m(T)} \tag{3.3}$$

where k ($k \in [0,1]$) is the CMSE parameter; l_0 is the initial matrix crack spacing; and $\sigma_{mocr}(T)$ is

$$\sigma_{mocr}(T) = \frac{E_m(T)}{E_c(T)} \sigma_{cr}(T) + E_m(T)[\alpha_c^{axial}(T) - \alpha_m^{axial}(T)]\Delta T \tag{3.4}$$

where $\alpha_c^{axial}(T)$ and $\alpha_m^{axial}(T)$ denote the temperature-dependent composite and matrix axial thermal expansion coefficient, respectively; ΔT denotes the temperature difference between the testing and fabricated temperature; and $\sigma_{cr}(T)$ denotes the temperature-dependent critical stress corresponding to composite's proportional limit stress.

$$\sigma_{cr}(T) = \left[\frac{6V_f^2 E_f(T) E_c^2(T) \tau_i(T) \Gamma_m(T)}{r_f V_m E_m^2(T)} \right]^{\frac{1}{3}}$$
$$- E_c(T)[\alpha_c^{axial}(T) - \alpha_m^{axial}(T)]\Delta T \tag{3.5}$$

where $\Gamma_m(T)$ denotes the temperature-dependent matrix fracture energy.

The matrix multiple cracking evolution can be determined using the following equation:

$$U_m(\sigma > \sigma_{cr}, T) = U_{crm}(\sigma_{cr}, T) \tag{3.6}$$

Experimental Comparisons

Experimental and theoretical predicted matrix cracking densities of SiC/SiC composites at room temperature and high temperatures of $T = 773$, 873, 973, and 1073 K were predicted, as shown in Figure 3.1.

- At room temperature, the matrix multiple cracking evolution occurs from $\sigma_{mc} = 240$ MPa with matrix cracking density $\lambda_m = 1.1$/mm to saturation at $\sigma_{sat} = 320$ MPa with $\lambda_m = 13$/mm.

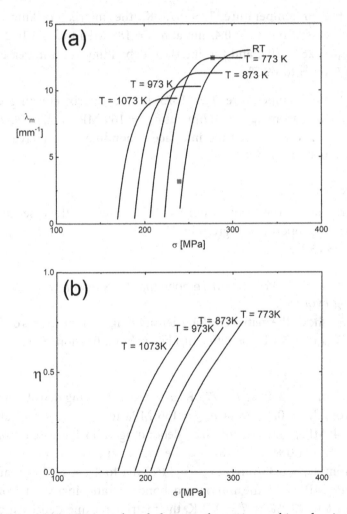

FIGURE 3.1 (a) Experimental and theoretical matrix cracking density versus applied stress curves; (b) the fiber/matrix interface debonding ratio versus applied stress curves of SiC/SiC composites.

- At high temperature T = 773 K, the matrix cracking density increases from λ_m = 0.5/mm at σ_{mc} = 222 MPa to λ_m = 12.4/mm at σ_{sat} = 311 MPa, and the interface debonding ratio ($\eta = 2l_d/l_c$) increases from η = 0.7% to 75.4%.

- At high temperature T = 873 K, the matrix cracking density increases from λ_m = 0.45/mm at σ_{mc} = 206 MPa to λ_m = 11.2/mm at σ_{sat} = 288 MPa, and the interface debonding ratio increases from η = 0.7% to 72.3%.

- At high temperature T = 973 K, the matrix cracking density increases from λ_m = 0.4/mm at σ_{mc} = 188 MPa to λ_m = 10.2/mm at σ_{sat} = 263 MPa, and the interface debonding ratio increases from η = 0.8% to 69.5%.

- At high temperature T = 1073 K, the matrix cracking density increases from λ_m = 0.35/mm at σ_{mc} = 169 MPa to λ_m = 9.4/mm at σ_{sat} = 236 MPa, and the interface debonding ratio increases from η = 0.8% to 66.9%.

Discussions

The ceramic composite system of SiC/SiC is used for the case study and its material properties are given by V_f = 30%, r_f = 7.5 μm, E_f = 230 GPa, and Γ_m = 15 J/m^2.

Effect of the Fiber Volume on Temperature-Dependent Matrix Multiple Cracking Evolution
Figure 3.2 shows the matrix cracking density at high temperatures of T = 773, 873, 973, and 1073 K for different fiber volume fractions (i.e., V_f = 30% and 35%).

- When V_f = 30% at T = 773 K, the matrix cracking density increased from λ_m = 0.4/mm at σ_{mc} = 156 MPa to λ_m = 13.7/mm at σ_{sat} = 234 MPa, and the interface debonding ratio increased from η = 0.7% to 1.0%; at T = 873 K, the matrix cracking density increased from λ_m = 0.35/mm at σ_{mc} = 146 MPa to λ_m = 11.8/mm at σ_{sat} = 219 MPa, and the interface debonding ratio increased from η = 0.8% to 97.2%; at T = 973 K, the matrix cracking density increased from λ_m = 0.3/mm at σ_{mc} = 134 MPa to λ_m = 10.2/mm at σ_{sat} = 202 MPa, and the interface debonding ratio increased from η = 0.9%

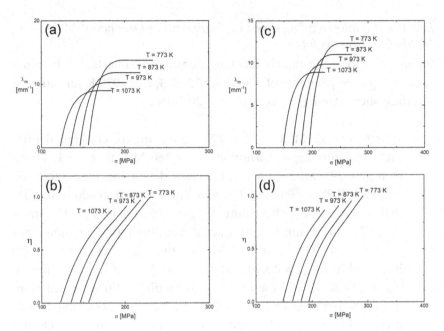

FIGURE 3.2 (a) Matrix cracking density at $V_f = 30\%$, (b) interface debonding ratio at $V_f = 30\%$, (c) matrix cracking density at $V_f = 35\%$, and (d) interface debonding ratio at $V_f = 35\%$ of SiC/SiC composites at high temperatures of $T = 773, 873, 973,$ and 1073 K.

to 91.9%; and at $T = 1073$ K, the matrix cracking density increased from $\lambda_m = 0.26$/mm at $\sigma_{mc} = 122$ MPa to $\lambda_m = 8.9$/mm at $\sigma_{sat} = 183$ MPa, and the interface debonding ratio increased from $\eta = 0.9\%$ to 87.2%.

- When $V_f = 35\%$ at $T = 773$ K, the matrix cracking density increased from $\lambda_m = 0.36$/mm at $\sigma_{mc} = 194$ MPa to $\lambda_m = 12.3$/mm at $\sigma_{sat} = 291$ MPa, and the interface debonding ratio increased from $\eta = 0.8\%$ to 1.0%; at $T = 873$ K, the matrix cracking density increased from $\lambda_m = 0.32$/mm at $\sigma_{mc} = 180$ MPa to $\lambda_m = 10.9$/mm at $\sigma_{sat} = 270$ MPa, and the interface debonding ratio increased from $\eta = 0.9\%$ to 95.2%; at $T = 973$ K, the matrix cracking density increased from $\lambda_m = 0.28$/mm at $\sigma_{mc} = 165$ MPa to $\lambda_m = 9.8$/mm at $\sigma_{sat} = 247$ MPa, and the interface debonding ratio increased from $\eta = 0.9\%$ to 90.8%; and at $T = 1073$ K, the matrix cracking density increased from $\lambda_m = 0.25$/mm at $\sigma_{mc} = 148$ MPa to $\lambda_m = 8.9$/mm at $\sigma_{sat} = 222$ MPa, and the interface debonding ratio increased from $\eta = 0.9\%$ to 86.7%.

Effect of the Interface Properties on Temperature-Dependent Multiple Matrix Cracking Evolution

Figure 3.3 shows the matrix cracking density and interface debonding ratio at high temperatures of $T = 773$, 873, 973, and 1073 K for different interface shear stress (i.e., $\tau_0 = 15$ and 20 MPa).

- When $\tau_0 = 15$ MPa at $T = 773$ K, the matrix cracking density increased from $\lambda_m = 0.4$/mm at $\sigma_{mc} = 166$ MPa to $\lambda_m = 13.7$/mm at $\sigma_{sat} = 249$ MPa, and the interface debonding ratio increased from $\eta = 0.8\%$ to 98%; at $T = 873$ K, the matrix cracking density increased from $\lambda_m = 0.34$/mm at $\sigma_{mc} = 156$ MPa to $\lambda_m = 12$/mm at $\sigma_{sat} = 234$ MPa, and the interface debonding ratio increased from $\eta = 0.9\%$ to 93.6%; at $T = 973$ K, the matrix cracking density increased from $\lambda_m = 0.3$/mm at $\sigma_{mc} = 145$ MPa to $\lambda_m = 10.7$/mm at $\sigma_{sat} = 217$ MPa, and the interface debonding ratio increased from $\eta = 0.9\%$ to 89.2%; and at $T = 1073$ K, the matrix cracking density increased from $\lambda_m = 0.27$/mm at $\sigma_{mc} = 132$ MPa to $\lambda_m = 9.6$/mm

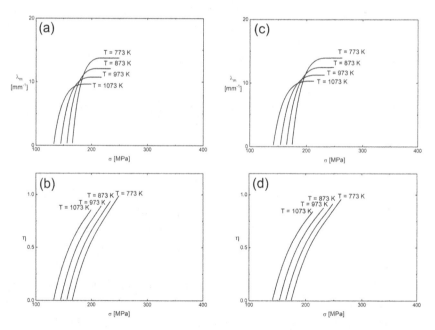

FIGURE 3.3 (a) Matrix cracking density at $\tau_0 = 15$ MPa, (b) interface debonding ratio at $\tau_0 = 15$ MPa, (c) matrix cracking density at $\tau_0 = 20$ MPa, and (d) interface debonding ratio at $\tau_0 = 20$ MPa of SiC/SiC composites at high temperatures of $T = 773$, 873, 973, and 1073 K.

at σ_{sat} = 199 MPa, and the interface debonding ratio increased from η = 0.9% to 85.4%.

- When τ_0 = 20 MPa at T = 773 K, the matrix cracking density increased from λ_m = 0.38/mm at σ_{mc} = 175 MPa to λ_m = 13.9/mm at σ_{sat} = 262 MPa, and the interface debonding ratio increased from η = 0.8% to 95%; at T = 873 K, the matrix cracking density increased from λ_m = 0.34/mm at σ_{mc} = 165 MPa to λ_m = 12.4/mm at σ_{sat} = 247 MPa, and the interface debonding ratio increased from η = 0.92% to 91.3%; at T = 973 K, the matrix cracking density increased from λ_m = 0.31/mm at σ_{mc} = 154 MPa to λ_m = 11.2/mm at σ_{sat} = 231 MPa, and the interface debonding ratio increased from η = 0.96% to 87.5%; and at T = 1073 K, the matrix cracking density increased from λ_m = 0.28/mm at σ_{mc} = 141 MPa to λ_m = 10.3/mm at σ_{sat} = 212 MPa, and the interface debonding ratio increased from η = 1.0% to 84.2%.

Figure 3.4 shows the matrix cracking density at high temperatures of T = 773, 873, 973, and 1073 K for different interface frictional coefficients (i.e., μ = 0.15 and 0.2).

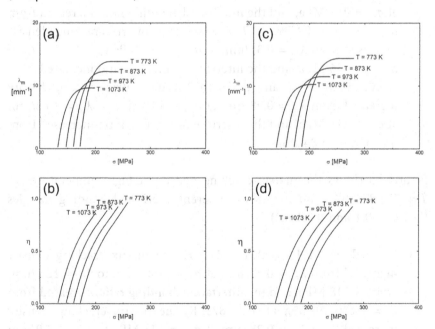

FIGURE 3.4 (a) Matrix cracking density at μ = 0.15, (b) interface debonding ratio at μ = 0.15, (c) matrix cracking density at μ = 0.2, and (d) interface debonding ratio at μ = 0.2 of SiC/SiC composites at high temperatures of T = 773, 873, 973, and 1073 K.

- When $\mu = 0.15$ at $T = 773$ K, the matrix cracking density increased from $\lambda_m = 0.38$/mm at $\sigma_{mc} = 172$ MPa to $\lambda_m = 13.8$/mm at $\sigma_{sat} = 258$ MPa, and the interface debonding ratio increased from $\eta = 0.87\%$ to 96.4%; at $T = 873$ K, the matrix cracking density increased from $\lambda_m = 0.34$/mm at $\sigma_{mc} = 160$ MPa to $\lambda_m = 12.2$/mm at $\sigma_{sat} = 240$ MPa, and the interface debonding ratio increased from $\eta = 0.91\%$ to 92.4%; at $T = 973$ K, the matrix cracking density increased from $\lambda_m = 0.3$/mm at $\sigma_{mc} = 147$ MPa to $\lambda_m = 10.8$/mm at $\sigma_{sat} = 220$ MPa, and the interface debonding ratio increased from $\eta = 0.95\%$ to 88.8%; and at $T = 1073$ K, the matrix cracking density increased from $\lambda_m = 0.27$/mm at $\sigma_{mc} = 132$ MPa to $\lambda_m = 9.6$/mm at $\sigma_{sat} = 198$ MPa, and the interface debonding ratio increased from $\eta = 0.9\%$ to 85.4%.

- When $\mu = 0.2$ at $T = 773$ K, the matrix cracking density increased from $\lambda_m = 0.38$/mm at $\sigma_{mc} = 186$ MPa to $\lambda_m = 14.4$/mm at $\sigma_{sat} = 279$ MPa, and the interface debonding ratio increased from $\eta = 0.9\%$ to 92.6%; at $T = 873$ K, the matrix cracking density increased from $\lambda_m = 0.35$/mm at $\sigma_{mc} = 172$ MPa to $\lambda_m = 12.8$/mm at $\sigma_{sat} = 259$ MPa, and the interface debonding ratio increased from $\eta = 0.94\%$ to 89.7%; at $T = 973$ K, the matrix cracking density increased from $\lambda_m = 0.31$/mm at $\sigma_{mc} = 157$ MPa to $\lambda_m = 11.5$/mm at $\sigma_{sat} = 236$ MPa, and the interface debonding ratio increased from $\eta = 0.97\%$ to 86.9%; and at $T = 1073$ K, the matrix cracking density increased from $\lambda_m = 0.28$/mm at $\sigma_{mc} = 141$ MPa to $\lambda_m = 10.2$/mm at $\sigma_{sat} = 211$ MPa, and the interface debonding ratio increased from $\eta = 0.99\%$ to 84.2%.

Figure 3.5 shows the matrix cracking density at high temperatures of $T = 773$, 873, 973, and 1073 K for different interface debonding energies (i.e., $\Gamma_i = 0.1$ and 0.5 J/m^2).

- When $\Gamma_i = 0.1$ J/m^2 at $T = 773$ K, the matrix cracking density increased from $\lambda_m = 0.32$/mm at $\sigma_{mc} = 156$ MPa to $\lambda_m = 12.6$/mm at $\sigma_{sat} = 218$ MPa, and the interface debonding ratio increased from $\eta = 1\%$ to 99.5%; at $T = 873$ K, the matrix cracking density increased from $\lambda_m = 0.28$/mm at $\sigma_{mc} = 146$ MPa to $\lambda_m = 10.9$/mm at $\sigma_{sat} = 204$ MPa, and the interface debonding ratio increased from $\lambda_m = 1.1\%$ to 94.7%; at $T = 973$ K, the matrix cracking density

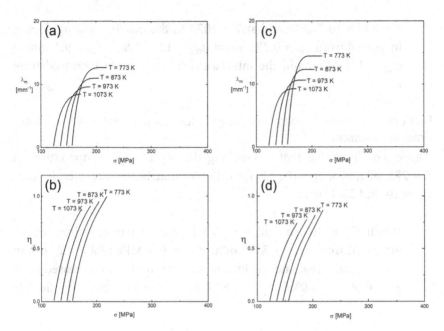

FIGURE 3.5 (a) Matrix cracking density at $\Gamma_i = 0.1$ J/m^2, (b) interface debonding ratio at $\Gamma_i = 0.1$ J/m^2, (c) matrix cracking density at $\Gamma_i = 0.5$ J/m^2, and (d) interface debonding ratio at $\Gamma_i = 0.5$ J/m^2 of SiC/SiC composites at high temperatures of $T = 773$, 873, 973, and 1073 K.

increased from $\lambda_m = 0.24$/mm at $\sigma_{mc} = 134$ MPa to $\lambda_m = 9.6$/mm at $\sigma_{sat} = 188$ MPa, and the interface debonding ratio increased from $\eta = 1.1\%$ to 90.6%; and at $T = 1073$ K, the matrix cracking density increased from $\lambda_m = 0.21$/mm at $\sigma_{mc} = 122$ MPa to $\lambda_m = 8.4$/mm at $\sigma_{sat} = 171$ MPa, and the interface debonding ratio increased from $\eta = 1.1\%$ to 86.8%.

- When $\Gamma_i = 0.5$ J/m^2 at $T = 773$ K, the matrix cracking density increased from $\lambda_m = 0.46$/mm at $\sigma_{mc} = 156$ MPa to $\lambda_m = 14.4$/mm at $\sigma_{sat} = 218$ MPa, and the interface debonding ratio increased from $\eta = 0.7\%$ to 86.4%; at $T = 873$ K, the matrix cracking density increased from $\lambda_m = 0.38$/mm at $\sigma_{mc} = 146$ MPa to $\lambda_m = 12.3$/mm at $\sigma_{sat} = 204$ MPa, and the interface debonding ratio increased from $\eta = 0.8\%$ to 81.6%; at $T = 973$ K, the matrix cracking density increased from $\lambda_m = 0.32$/mm at $\sigma_{mc} = 134$ MPa to $\lambda_m = 10.6$/mm at $\sigma_{sat} = 188$ MPa, and the interface debonding ratio increased from

η = 0.85% to 77.6%; and at T = 1073 K, the matrix cracking density increased from λ_m = 0.28/mm at σ_{mc} = 122 MPa to λ_m = 9.2/mm at σ_{sat} = 171 MPa, and the interface debonding ratio increased from η = 0.9% to 74.2%.

Effect of the Matrix Property on Temperature-Dependent Multiple Matrix Cracking Evolution

Figure 3.6 shows the matrix cracking density at high temperatures of T = 773, 873, 973, and 1073 K for different matrix fracture energies (i.e., Γ_m = 10 and 20 J/m^2).

- When Γ_m = 10 J/m^2 at T = 773 K, the matrix cracking density increased from λ_m = 0.83/mm at σ_{mc} = 134 MPa to λ_m = 23.6/mm at σ_{sat} = 201 MPa, and the interface debonding ratio increased from η = 0.3% to 1.0%; at T = 873 K, the matrix cracking density

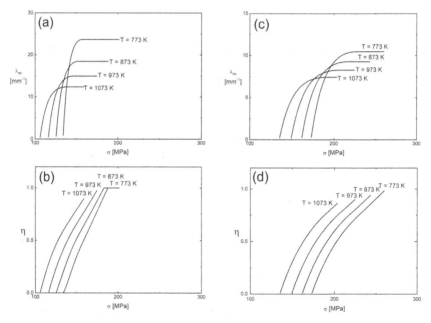

FIGURE 3.6 (a) Matrix cracking density at Γ_m = 10 J/m^2, (b) interface debonding ratio at Γ_m = 10 J/m^2, (c) matrix cracking density at Γ_m = 20 J/m^2, and (d) interface debonding ratio at Γ_m = 20 J/m^2 of SiC/SiC composites at T = 773, 873, 973, and 1073 K.

increased from $\lambda_m = 0.62$/mm at $\sigma_{mc} = 125$ MPa to $\lambda_m = 18.4$/mm at $\sigma_{sat} = 188$ MPa, and the interface debonding ratio increased from $\eta = 0.5\%$ to 1.0%; at $T = 973$ K, the matrix cracking density increased from $\lambda_m = 0.48$/mm at $\sigma_{mc} = 116$ MPa to $\lambda_m = 14.9$/mm at $\sigma_{sat} = 174$ MPa, and the interface debonding ratio increased from $\eta = 0.69\%$ to 97.4%; and at $T = 1073$ K, the matrix cracking density increased from $\lambda_m = 0.38$/mm at $\sigma_{mc} = 106$ MPa to $\lambda_m = 12.3$/mm at $\sigma_{sat} = 158$ MPa, and the interface debonding ratio increased from $\eta = 0.79\%$ to 89.7%.

- When $\Gamma_m = 20$ J/m^2 at $T = 773$ K, the matrix cracking density increased from $\lambda_m = 0.29$/mm at $\sigma_{mc} = 173$ MPa to $\lambda_m = 10.4$/mm at $\sigma_{sat} = 260$ MPa, and the interface debonding ratio increased from $\eta = 0.9\%$ to 98.3%; at $T = 873$ K, the matrix cracking density increased from $\lambda_m = 0.26$/mm at $\sigma_{mc} = 162$ MPa to $\lambda_m = 9.2$/mm at $\sigma_{sat} = 243$ MPa, and the interface debonding ratio increased from $\eta = 0.98\%$ to 94%; at $T = 973$ K, the matrix cracking density increased from $\lambda_m = 0.23$/mm at $\sigma_{mc} = 149$ MPa to $\lambda_m = 8.2$/mm at $\sigma_{sat} = 224$ MPa, and the interface debonding ratio increased from $\eta = 1\%$ to 90%; and at $T = 1073$ K, the matrix cracking density increased from $\lambda_m = 0.2$/mm at $\sigma_{mc} = 135$ MPa to $\lambda_m = 7.4$/mm at $\sigma_{sat} = 203$ MPa, and the interface debonding ratio increased from $\eta = 1.0\%$ to 86.4%.

TIME-DEPENDENT MULTIPLE MATRIX CRACKING EVOLUTION IN SiC/SiC COMPOSITES

In this section, the time-dependent multiple matrix cracking of SiC/SiC composites was investigated considering the damage mechanism of interface oxidation. Experimental matrix cracking density of SiC/SiC composites at high temperatures was predicted. Effects of fiber volume fraction, interface properties, and matrix properties on the multiple matrix cracking density of SiC/SiC composite at different temperatures were also discussed.

Micromechanical Model

For interface partial debonding, the time/temperature-dependent matrix strain energy $U_m(t, T)$ can be described using the following equation:

$$U_m(t, T) = \frac{A_m}{E_m}\left\{\frac{4}{3}\left(\frac{V_f}{V_m}\frac{\tau_f(T)}{r_f(T)}\right)^2\zeta^3(t, T) + 4\left(\frac{V_f}{V_m}\frac{\tau_f(T)}{r_f}\right)^2[l_d(t, T)\right.$$

$$- \zeta(t, T)]\zeta^2(t, T)$$

$$+ 4\left(\frac{V_f}{r_f V_m}\right)^2\tau_f(T)\tau_i(T)\zeta(t, T)[l_d(t, T) - \zeta(t, T)]^2$$

$$+ \frac{4}{3}\left[\frac{V_f}{V_m}\frac{\tau_i(T)}{r_f}\right]^2[l_d(t, T) - \zeta(t, T)]^3 + \sigma_{mo}^2\left[\frac{l_c(t, T)}{2} - l_d(t, T)\right]$$

$$+ \frac{2r_f\sigma_{mo}(T)}{\rho}\left[2\frac{V_f}{V_m}\frac{\tau_f(T)}{r_f}\zeta(t, T) + 2\frac{V_f}{V_m}\frac{\tau_i(T)}{r_f}[l_d(t, T) - \zeta(t, T)]\right.$$

$$- \sigma_{mo}(T)\Big]$$

$$\times\left[1 - \exp\left(-\rho\frac{l_c(t, T)/2 - l_d(t, T)}{r_f}\right)\right]$$

$$+ \frac{r_f}{2\rho}\left[2\frac{V_f}{V_m}\frac{\tau_f(T)}{r_f}\zeta(t, T) + 2\frac{V_f}{V_m}\frac{\tau_i(T)}{r_f}(l_d(t, T) - \zeta(t, T)) - \sigma_{mo}(T)\right]^2$$

$$\times\left.\left[1 - \exp\left(-2\rho\frac{l_c(t, T)/2 - l_d(t, T)}{r_f}\right)\right]\right\}$$

$$(3.7)$$

where $\zeta(t, T)$ denotes the time/temperature-dependent interface oxidation length.

For interface complete debonding, the time/temperature-dependent matrix strain energy $U_m(t, T)$ can be described using the following equation:

$$U_m(t, T) = \frac{A_m}{E_m}\left\{\frac{4}{3}\left[\frac{V_f}{V_m}\frac{\tau_f(T)}{r_f}\right]^2\zeta^3(t, T) + 4\left[\frac{V_f}{V_m}\frac{\tau_f(T)}{r_f}\right]^2[l_d(t, T)\right.$$

$$- \zeta(t, T)]\zeta^2(t, T)$$

$$+ 4\left(\frac{V_f}{r_f V_m}\right)^2\tau_f(T)\tau_i(T)\zeta(t, T)[l_d(t, T) - \zeta(t, T)]^2$$

$$+ \left.\frac{4}{3}\left[\frac{V_f}{V_m}\frac{\tau_i(T)}{r_f}\right]^2[l_d(t, T) - \zeta(t, T)]^3\right\}$$

$$(3.8)$$

The CMSE $U_m^{cr}(T)$ can be described using the following equation:

$$U_m^{cr}(T) = \frac{1}{2}kA_m l_0\frac{\sigma_{mocr}^2(T)}{E_m(T)} \qquad (3.9)$$

Experimental Comparisons

Zhang et al. [34] investigated the damage evolution of multiple matrix cracking in SiC/SiC composites. Evolution of matrix multiple cracking at different temperatures was analyzed, as shown in Figure 3.7.

- At room temperature, the multiple matrix cracking evolution started from $\sigma_{mc} = 135$ MPa and approached saturation at $\sigma_{sat} = 250$ MPa, and matrix multiple cracking density increased from $\lambda_m = 0.4$/mm to 2.4/mm.

- At $T = 973$ K and $t = 3$ h, the matrix cracking density increased from $\lambda_m = 0.06$/mm at $\sigma_{mc} = 112$ MPa to $\lambda_m = 2.0$/mm at $\sigma_{sat} = 140$ MPa, and the interface oxidation ratio ($y = \zeta/l_d$) decreased from $y = 9.8\%$ to 4.6%.

- At $T = 1073$ K and $t = 3$ h, the matrix cracking density increased from $\lambda_m = 0.04$/mm at $\sigma_{mc} = 98.4$ MPa to $\lambda_m = 1.9$/mm at $\sigma_{sat} = 136$ MPa, and the interface oxidation ratio decreased from $y = 23.7\%$ to 11.6%.

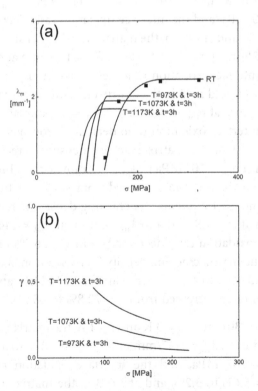

FIGURE 3.7 (a) Experimental and theoretical matrix cracking density; (b) interface oxidation ratio of SiC/SiC composites.

- At $T = 1173$ K and $t = 3$ h, the matrix cracking density increased from $\lambda_m = 0.03$/mm at $\sigma_{mc} = 83.4$ MPa to $\lambda_m = 1.7$/mm at $\sigma_{sat} = 133$ MPa, and the interface oxidation ratio decreased from $\gamma = 46.2\%$ to 24.5%.

Discussions

In this section, the effects of fiber volume fraction, interface properties, and matrix properties on the time-dependent matrix multiple cracking evolution in SiC/SiC composites were discussed.

Effect of the Fiber Volume on Time-Dependent Matrix Multiple Cracking Evolution

Figure 3.8 shows the effect of fiber volume fraction (i.e., $V_f = 25\%$, 30%, and 35%) on the time-dependent matrix multiple cracking of SiC/SiC composites at $T = 873, 973$, and 1073 K for the oxidation duration of $t = 1$ and 3 h.

- When $V_f = 25\%$ at $T = 873$ K and $t = 1$ h, the matrix cracking density increased from $\lambda_m = 1.2$/mm at $\sigma_{mc} = 110$ MPa to $\lambda_m = 11.9$/mm at $\sigma_{sat} = 120$ MPa, and the interface oxidation ratio decreased from $\gamma = 6\%$ to 3.4%; and at $t = 3$ h, the matrix cracking density increased from $\lambda_m = 0.67$/mm at $\sigma_{mc} = 110$ MPa to $\lambda_m = 10.7$/mm at $\sigma_{sat} = 123$ MPa, and the interface oxidation ratio decreased from $\gamma = 16.6\%$ to 9.8%. At $T = 973$ K and $t = 1$ h, the matrix cracking density increased from $\lambda_m = 0.24$/mm at $\sigma_{mc} = 104$ MPa to $\lambda_m = 9.2$/mm at $\sigma_{sat} = 133$ MPa, and the interface oxidation ratio decreased from $\gamma = 16.7\%$ to 9.5%; and at $t = 3$ h, the matrix cracking density increased from $\lambda_m = 0.17$/mm at $\sigma_{mc} = 104$ MPa to $\lambda_m = 7.4$/mm at $\sigma_{sat} = 140$ MPa, and the interface oxidation ratio decreased from $\gamma = 40.5\%$ to 25.1%. At $T = 1073$ K and $t = 1$ h, the matrix cracking density increased from $\lambda_m = 0.15$/mm at $\sigma_{mc} = 96$ MPa to $\lambda_m = 7$/mm at $\sigma_{sat} = 144$ MPa, and the interface oxidation ratio decreased from $\gamma = 36.7\%$ to 21.1%; and at $t = 3$ h, the matrix cracking density increased from $\lambda_m = 0.09$/mm at $\sigma_{mc} = 96.3$ MPa to $\lambda_m = 5.0$/mm at $\sigma_{sat} = 144$ MPa, and the interface oxidation ratio decreased from $\gamma = 72.8\%$ to 48.9%.

- When $V_f = 30\%$ at $T = 873$ K and $t = 1$ h, the matrix cracking density increased from $\lambda_m = 0.47$/mm at $\sigma_{mc} = 146$ MPa to $\lambda_m = 10.36$/mm at $\sigma_{sat} = 167$ MPa, and the interface oxidation ratio decreased from $\gamma = 5.6\%$ to 3.2%; and at $t = 3$ h, the matrix cracking density increased from $\lambda_m = 0.35$/mm at $\sigma_{mc} = 146$ MPa to $\lambda_m = 3.4$/mm at

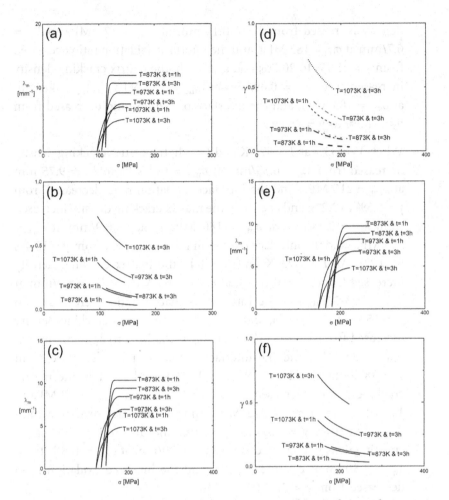

FIGURE 3.8 (a) Matrix cracking density at $V_f = 25\%$, (b) interface oxidation ratio at $V_f = 25\%$, (c) matrix cracking density at $V_f = 30\%$, (d) interface oxidation ratio at $V_f = 30\%$, (e) matrix cracking density at $V_f = 35\%$, and (f) interface oxidation ratio at $V_f = 35\%$.

$\sigma_{sat} = 170$ MPa, and the interface oxidation ratio decreased from $\gamma = 15.6\%$ to 9.3%. At $T = 973$ K and $t = 1$ h, the matrix cracking density increased from $\lambda_m = 0.21$/mm at $\sigma_{mc} = 134$ MPa to $\lambda_m = 8.46$/mm at $\sigma_{sat} = 178$ MPa, and the interface oxidation ratio decreased from $\gamma = 15.9\%$ to 9.1%; and at $t = 3$ h, the matrix cracking density increased from $\lambda_m = 0.15$/mm at $\sigma_{mc} = 134$ MPa to $\lambda_m = 6.9$/mm at $\sigma_{sat} = 186$ MPa, and the interface oxidation ratio decreased from $\gamma = 38.9\%$ to 24.3%. At $T = 1073$ K and $t = 1$ h, the matrix cracking density increased from $\lambda_m = 0.14$/mm at $\sigma_{mc} = 122$ MPa to $\lambda_m = 6.7$/mm at $\sigma_{sat} = 183$ MPa, and the interface oxidation ratio decreased from $\gamma = 35.7\%$ to 20.7%; and at $t = 3$ h, the matrix cracking density increased from $\lambda_m = 0.09$/mm at $\sigma_{mc} = 122$ MPa to $\lambda_m = 4.8$/mm at $\sigma_{sat} = 183$ MPa, and the interface oxidation ratio decreased from $\gamma = 71.5\%$ to $\gamma = 48.2\%$.

- When $V_f = 35\%$ at $T = 873$ K and $t = 1$ h, the matrix cracking density increased from $\lambda_m = 0.35$/mm at $\sigma_{mc} = 180$ MPa to $\lambda_m = 9.75$/mm at $\sigma_{sat} = 215$ MPa, and the interface oxidation ratio decreased from $\gamma = 5.5\%$ to 3.2%; and at $t = 3$ h, the matrix cracking density increased from $\lambda_m = 0.28$/mm at $\sigma_{mc} = 180$ MPa to $\lambda_m = 8.9$/mm at $\sigma_{sat} = 219$ MPa, and the interface oxidation ratio decreased from $\gamma = 15.4\%$ to 9.2%. At $T = 973$ K and $t = 1$ h, the matrix cracking density increased from $\lambda_m = 0.2$/mm at $\sigma_{mc} = 165$ MPa to $\lambda_m = 8.2$/mm at $\sigma_{sat} = 223$ MPa, and the interface oxidation ratio decreased from $\gamma = 15.9\%$ to $\gamma = 9.2\%$; and at $t = 3$ h, the matrix cracking density increased from $\lambda_m = 0.15$/mm at $\sigma_{mc} = 165$ MPa to $\lambda_m = 6.7$/mm at $\sigma_{sat} = 234$ MPa, and the interface oxidation ratio decreased from $\gamma = 38.8\%$ to $\gamma = 24.3\%$. At $T = 1073$ K and $t = 1$ h, the matrix cracking density increased from $\lambda_m = 0.14$/mm at $\sigma_{mc} = 148$ MPa to $\lambda_m = 6.6$/mm at $\sigma_{sat} = 222$ MPa, and the interface oxidation ratio decreased from $\gamma = 36\%$ to $\gamma = 21\%$; and at $t = 3$ h, the matrix cracking density increased from $\lambda_m = 0.09$/mm at $\sigma_{mc} = 148$ MPa to $\lambda_m = 4.8$/mm at $\sigma_{sat} = 222$ MPa, and the interface oxidation ratio decreased from $\gamma = 71.9\%$ to 48.7%.

Effect of the Interface Properties on Time-Dependent Matrix Multiple Cracking Evolution

Figure 3.9 shows the effect of interface shear stress (i.e., $\tau_0 = 15$, 20, and 25 MPa) on the time-dependent matrix multiple cracking of SiC/SiC composites at $T = 873$, 973, and 1073 K for $t = 1$ and 3 h.

- When $\tau_0 = 15$ MPa at $T = 873$ K and $t = 1$ h, the matrix cracking density increased from $\lambda_m = 0.34$/mm at $\sigma_{mc} = 191$ MPa to $\lambda_m = 10.33$/mm at $\sigma_{sat} = 234$ MPa, and the interface oxidation ratio decreased from $\gamma = 6\%$ to 3.5%; and at $t = 3$ h, the matrix cracking density increased from $\lambda_m = 0.27$/mm at $\sigma_{mc} = 191$ MPa to $\lambda_m = 9.3$/mm at $\sigma_{sat} = 240$ MPa, and the interface oxidation ratio decreased

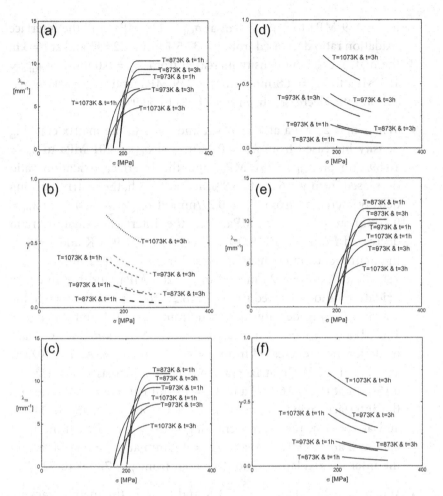

FIGURE 3.9   (a) Matrix cracking density at τ_0 = 15 MPa, (b) interface oxidation ratio at τ_0 = 15 MPa, (c) matrix cracking density at τ_0 = 20 MPa, (d) interface oxidation ratio at τ_0 = 20 MPa, (e) matrix cracking density at τ_0 = 25 MPa, and (f) interface oxidation ratio at τ_0 = 25 MPa.

from γ = 16.6% to 10.2%. At T = 973 K and t = 1 h, the matrix cracking density increased from λ_m = 0.21/mm at σ_{mc} = 176 MPa to λ_m = 8.7/mm at σ_{sat} = 245 MPa, and the interface oxidation ratio decreased from γ = 17.1% to γ = 10.1%; and at t = 3 h, the matrix cracking density increased from λ_m = 0.15/mm at σ_{mc} = 176 MPa to λ_m = 7/mm at σ_{sat} = 258 MPa, and the interface oxidation ratio decreased from γ = 40.6% to γ = 26.3%. At T = 1073 K and t = 1 h, the matrix cracking density increased from λ_m = 0.15/mm at

σ_{mc} = 159 MPa to λ_m = 7/mm at σ_{sat} = 239 MPa, and the interface oxidation ratio decreased from γ = 37.5% to γ = 22.8%; and at t = 3 h, the matrix cracking density increased from λ_m = 0.09/mm at σ_{mc} = 159 MPa to λ_m = 4.8/mm at σ_{sat} = 239 MPa, and the interface oxidation ratio decreased from γ = 71.7% to 50.9%.

- When τ_0 = 20 MPa at T = 873 K and t = 1 h, the matrix cracking density increased from λ_m = 0.34/mm at σ_{mc} = 201 MPa to λ_m = 10.89/mm at σ_{sat} = 251 MPa, and the interface oxidation ratio decreased from γ = 6.5% to 3.9%; and at t = 3 h, the matrix cracking density increased from λ_m = 0.27/mm at σ_{mc} = 201 MPa to λ_m = 9.7/mm at σ_{sat} = 260 MPa, and the interface oxidation ratio decreased from γ = 17.7% to γ = 11%. At T = 973 K and t = 1 h, the matrix cracking density increased from λ_m = 0.22/mm at σ_{mc} = 186 MPa to λ_m = 9.26/mm at σ_{sat} = 265 MPa, and the interface oxidation ratio decreased from γ = 18.2% to 11%; and at t = 3 h, the matrix cracking density increased from λ_m = 0.15/mm at σ_{mc} = 186 MPa to λ_m = 7.3/mm at σ_{sat} = 280 MPa, and the interface oxidation ratio decreased from γ = 42.3% to 28.1%. At T = 1073 K and t = 1 h, the matrix cracking density increased from λ_m = 0.15/mm at σ_{mc} = 169 MPa to λ_m = 7.4/mm at σ_{sat} = 254 MPa, and the interface oxidation ratio decreased from γ = 39% to 24.5%; and at t = 3 h, the matrix cracking density increased from λ_m = 0.09/mm at σ_{mc} = 169 MPa to λ_m = 4.9/mm at σ_{sat} = 254 MPa, and the interface oxidation ratio decreased from γ = 72.1% to 52.9%.

- When τ_0 = 25 MPa at T = 873 K and t = 1 h, the matrix cracking density increased from λ_m = 0.35/mm at σ_{mc} = 211 MPa to λ_m = 11.4/mm at σ_{sat} = 270 MPa, and the interface oxidation ratio decreased from γ = 7% to 4.2%; and at t = 3 h, the matrix cracking density increased from λ_m = 0.28/mm at σ_{mc} = 211 MPa to λ_m = 10.2/mm at σ_{sat} = 277 MPa, and the interface oxidation ratio decreased from γ = 18.8% to 11.8%. At T = 973 K and t = 1 h, the matrix cracking density increased from λ_m = 0.23/mm at σ_{mc} = 195 MPa to λ_m = 9.7/mm at σ_{sat} = 282 MPa, and the interface oxidation ratio decreased from γ = 19.3% to 11.8%; and at t = 3 h, the matrix cracking density increased from λ_m = 0.15/mm at σ_{mc} = 195 MPa to λ_m = 7.5/mm at σ_{sat} = 293 MPa, and the interface oxidation ratio decreased from γ = 43.8% to 29.7%. At T = 1073 K and t = 1 h, the matrix cracking density increased from λ_m = 0.16/mm at

σ_{mc} = 178 MPa to λ_m = 7.7/mm at σ_{sat} = 268 MPa, and the interface oxidation ratio decreased from γ = 40.4% to 26.1%; and at t = 3 h, the matrix cracking density increased from λ_m = 0.09/mm at σ_{mc} = 178 MPa to λ_m = 5.0/mm at σ_{sat} = 268 MPa, and the interface oxidation ratio decreased from γ = 72.6% to 54.6%.

Figure 3.10 shows the effect of interface shear stress (i.e., τ_f = 1, 3, and 5 MPa) on the time-dependent matrix cracking of SiC/SiC composites at T = 873, 973, and 1073 K for t = 1 and 3 h.

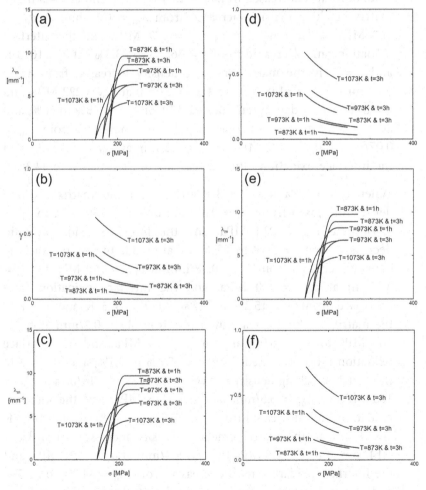

FIGURE 3.10 (a) Matrix cracking density at τ_f = 1 MPa, (b) interface oxidation ratio at τ_f = 1 MPa, (c) matrix cracking density at τ_f = 3 MPa, (d) interface oxidation ratio at τ_f = 3 MPa, (e) matrix cracking density at τ_f = 5 MPa, and (f) interface oxidation ratio at τ_f = 5 MPa.

- When $\tau_f = 1$ MPa at $T = 873$ K and $t = 1$ h, the matrix cracking density increased from $\lambda_m = 0.34$/mm at $\sigma_{mc} = 180$ MPa to $\lambda_m = 9.6$/mm at $\sigma_{sat} = 214$ MPa, and the interface oxidation ratio decreased from $\gamma = 5.5\%$ to 3.2%; and at $t = 3$ h, the matrix cracking density increased from $\lambda_m = 0.27$/mm at $\sigma_{mc} = 180$ MPa to $\lambda_m = 8.6$/mm at $\sigma_{sat} = 220$ MPa, and the interface oxidation ratio decreased from $\gamma = 15\%$ to 9.1%. At $T = 973$ K and $t = 1$ h, the matrix cracking density increased from $\lambda_m = 0.19$/mm at $\sigma_{mc} = 165$ MPa to $\lambda_m = 7.9$/mm at $\sigma_{sat} = 225$ MPa, and the interface oxidation ratio decreased from $\gamma = 15.4\%$ to 9%; and at $t = 3$ h, the matrix cracking density increased from $\lambda_m = 0.13$/mm at $\sigma_{mc} = 165$ MPa to $\lambda_m = 6.3$/mm at $\sigma_{sat} = 237$ MPa, and the interface oxidation ratio decreased from $\gamma = 36.2\%$ to 23.3%. At $T = 1073$ K and $t = 1$ h, the matrix cracking density increased from $\lambda_m = 0.13$/mm at $\sigma_{mc} = 148$ MPa to $\lambda_m = 6.3$/mm at $\sigma_{sat} = 222$ MPa, and the interface oxidation ratio decreased from $\gamma = 33.5\%$ to 20%; and at $t = 3$ h, the matrix cracking density increased from $\lambda_m = 0.07$/mm at $\sigma_{mc} = 148$ MPa to $\lambda_m = 4.2$/mm at $\sigma_{sat} = 222$ MPa, and the interface oxidation ratio decreased from $\gamma = 62.7\%$ to 44.3%.

- When $\tau_f = 3$ MPa at $T = 873$ K and $t = 1$ h, the matrix cracking density increased from $\lambda_m = 0.35$/mm at $\sigma_{mc} = 180$ MPa to $\lambda_m = 9.7$/mm at $\sigma_{sat} = 214$ MPa, and the interface oxidation ratio decreased from $\gamma = 5.5\%$ to 3.2%; and at $t = 3$ h, the matrix cracking density increased from $\lambda_m = 0.28$/mm at $\sigma_{mc} = 180$ MPa to $\lambda_m = 8.7$/mm at $\sigma_{sat} = 220$ MPa, and the interface oxidation ratio decreased from $\gamma = 15.2\%$ to 9.2%. At $T = 973$ K and $t = 1$ h, the matrix cracking density increased from $\lambda_m = 0.2$/mm at $\sigma_{mc} = 165$ MPa to $\lambda_m = 8.0$/mm at $\sigma_{sat} = 224$ MPa, and the interface oxidation ratio decreased from $\gamma = 15.6\%$ to 9.1%; and at $t = 3$ h, the matrix cracking density increased from $\lambda_m = 0.14$/mm at $\sigma_{mc} = 165$ MPa to $\lambda_m = 6.5$/mm at $\sigma_{sat} = 236$ MPa, and the interface oxidation ratio decreased from $\gamma = 37.5\%$ to 23.8%. At $T = 1073$ K and $t = 1$ h, the matrix cracking density increased from $\lambda_m = 0.14$/mm at $\sigma_{mc} = 148$ MPa to $\lambda_m = 6.4$/mm at $\sigma_{sat} = 222$ MPa, and the interface oxidation ratio decreased from $\lambda_m = 34.7\%$ to 20.5%; and at $t = 3$ h, the matrix cracking density increased from $\lambda_m = 0.08$/mm at $\sigma_{mc} = 148$ MPa to $\lambda_m = 4.5$/mm at $\sigma_{sat} = 222$ MPa, and the interface oxidation ratio decreased from $\gamma = 67\%$ to 46.4%.

- When $\tau_f = 5$ MPa at $T = 873$ K and $t = 1$ h, the matrix cracking density increased from $\lambda_m = 0.35$/mm at $\sigma_{mc} = 180$ MPa to $\lambda_m = 9.7$/mm at $\sigma_{sat} = 214$ MPa, and the interface oxidation ratio decreased from $\gamma = 5.5\%$ to 3.2%; and at $t = 3$ h, the matrix cracking density increased from $\lambda_m = 0.29$/mm at $\sigma_{mc} = 180$ MPa to $\lambda_m = 8.9$/mm at $\sigma_{sat} = 220$ MPa, and the interface oxidation ratio decreased from $\gamma = 15.4\%$ to 9.3%. At $T = 973$ K and $t = 1$ h, the matrix cracking density increased from $\lambda_m = 0.2$/mm at $\sigma_{mc} = 165$ MPa to $\lambda_m = 8.2$/mm at $\sigma_{sat} = 224$ MPa, and the interface oxidation ratio decreased from $\gamma = 15.9\%$ to 9.2%; and at $t = 3$ h, the matrix cracking density increased from $\lambda_m = 0.15$/mm at $\sigma_{mc} = 165$ MPa to $\lambda_m = 6.7$/mm at $\sigma_{sat} = 234$ MPa, and the interface oxidation ratio decreased from $\gamma = 38.8\%$ to 24.3%. At $T = 1073$ K and $t = 1$ h, the matrix cracking density increased from $\lambda_m = 0.15$/mm at $\sigma_{mc} = 148$ MPa to $\lambda_m = 6.6$/mm at $\sigma_{sat} = 222$ MPa, and the interface oxidation ratio decreased from $\gamma = 36\%$ to 21%; and at $t = 3$ h, the matrix cracking density increased from $\lambda_m = 0.09$/mm at $\sigma_{mc} = 148$ MPa to $\lambda_m = 4.8$/mm at $\sigma_{sat} = 222$ MPa, and the interface oxidation ratio decreased from $\gamma = 72\%$ to 48.7%.

Figure 3.11 shows the effect of interface frictional coefficient (i.e., $\mu = 0.05$, 0.1, and 0.15) on the time-dependent matrix cracking of SiC/SiC composites at $T = 873$, 973, and 1073 K for $t = 1$ and 3 h.

- When $\mu = 0.05$ at $T = 873$ K and $t = 1$ h, the matrix cracking density increased from $\lambda_m = 1.2$/mm at $\sigma_{mc} = 128$ MPa to $\lambda_m = 9.9$/mm at $\sigma_{sat} = 136$ MPa, and the interface oxidation ratio decreased from $\gamma = 4.9\%$ to 2.7%; and at $t = 3$ h, the matrix cracking density increased from $\lambda_m = 0.71$/mm at $\sigma_{mc} = 128$ MPa to $\lambda_m = 9.2$/mm at $\sigma_{sat} = 138$ MPa, and the interface oxidation ratio decreased from $\gamma = 13.8\%$ to 7.9%. At $T = 973$ K and $t = 1$ h, the matrix cracking density increased from $\lambda_m = 0.21$/mm at $\sigma_{mc} = 120$ MPa to $\lambda_m = 7.9$/mm at $\sigma_{sat} = 151$ MPa, and the interface oxidation ratio decreased from $\gamma = 14.5\%$ to 7.9%; and at $t = 3$ h, the matrix cracking density increased from $\lambda_m = 0.16$/mm at $\sigma_{mc} = 120$ MPa to $\lambda_m = 6.7$/mm at $\sigma_{sat} = 156$ MPa, and the interface oxidation ratio decreased from $\gamma = 37.1\%$ to 21.7%. At $T = 1073$ K and $t = 1$ h, the matrix cracking density increased from $\lambda_m = 0.14$/mm at $\sigma_{mc} = 110$ MPa to $\lambda_m = 6.3$/mm at $\sigma_{sat} = 162$ MPa, and the interface

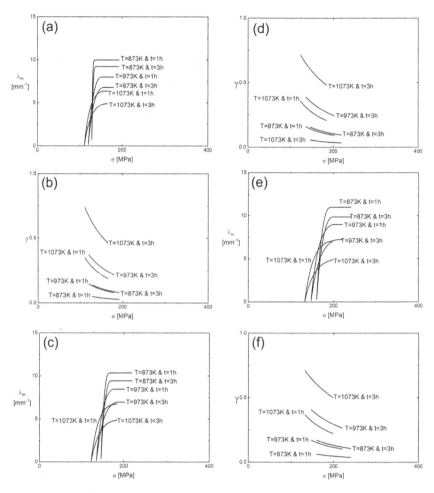

FIGURE 3.11 (a) Matrix cracking density at $\mu = 0.05$, (b) interface oxidation ratio at $\mu = 0.05$, (c) matrix cracking density at $\mu = 0.1$, (d) interface oxidation ratio at $\mu = 0.1$, (e) matrix cracking density at $\mu = 0.15$, and (f) interface oxidation ratio at $\mu = 0.15$.

oxidation ratio decreased from $\gamma = 34.8\%$ to 18.8%; and at $t = 3$ h, the matrix cracking density increased from $\lambda_m = 0.09$/mm at $\sigma_{mc} = 110$ MPa to $\lambda_m = 4.8$/mm at $\sigma_{sat} = 166$ MPa, and the interface oxidation ratio decreased from $\gamma = 73.7\%$ to 46.1%.

- When $\mu = 0.1$ at $T = 873$ K and $t = 1$ h, the matrix cracking density increased from $\lambda_m = 0.47$/mm at $\sigma_{mc} = 146$ MPa to $\lambda_m = 10.3$/mm at $\sigma_{sat} = 167$ MPa, and the interface oxidation ratio decreased from $\gamma = 5.6\%$ to 3.2%; and at $t = 3$ h, the matrix cracking density increased from $\lambda_m = 0.35$/mm at $\sigma_{mc} = 146$ MPa to $\lambda_m = 9.4$/mm at $\sigma_{sat} = 170$ MPa, and the interface oxidation ratio decreased from $\gamma = 15.6\%$ to 9.3%. At $T = 973$ K and $t = 1$ h, the matrix cracking density increased from $\lambda_m = 0.21$/mm at $\sigma_{mc} = 134$ MPa to $\lambda_m = 8.4$/mm at $\sigma_{sat} = 178$ MPa, and the interface oxidation ratio decreased from $\gamma = 15.9\%$ to 9.1%; and at $t = 3$ h, the matrix cracking density increased from $\lambda_m = 0.15$/mm at $\sigma_{mc} = 134$ MPa to $\lambda_m = 6.9$/mm at $\sigma_{sat} = 186$ MPa, and the interface oxidation ratio decreased from $\gamma = 38.9\%$ to 24.3%. At $T = 1073$ K and $t = 1$ h, the matrix cracking density increased from $\lambda_m = 0.14$/mm at $\sigma_{mc} = 122$ MPa to $\lambda_m = 6.7$/mm at $\sigma_{sat} = 183$ MPa, and the interface oxidation ratio decreased from $\gamma = 35.7\%$ to 20.7%; and at $t = 3$ h, the matrix cracking density increased from $\lambda_m = 0.09$/mm at $\sigma_{mc} = 122$ MPa to $\lambda_m = 4.8$/mm at $\sigma_{sat} = 183$ MPa, and the interface oxidation ratio decreased from $\gamma = 71.5\%$ to 48.2%.

• When $\mu = 0.15$ at $T = 873$ K and $t = 1$ h, the matrix cracking density increased from $\lambda_m = 0.4$/mm at $\sigma_{mc} = 160$ MPa to $\lambda_m = 10.9$/mm at $\sigma_{sat} = 193$ MPa, and the interface oxidation ratio decreased from $\gamma = 6.3\%$ to 3.7%; and at $t = 3$ h, the matrix cracking density increased from $\lambda_m = 0.31$/mm at $\sigma_{mc} = 160$ MPa to $\lambda_m = 9.8$/mm at $\sigma_{sat} = 198$ MPa, and the interface oxidation ratio decreased from $\gamma = 17.1\%$ to 10.5%. At $T = 973$ K and $t = 1$ h, the matrix cracking density increased from $\lambda_m = 0.22$/mm at $\sigma_{mc} = 147$ MPa to $\lambda_m = 8.9$/mm at $\sigma_{sat} = 201$ MPa, and the interface oxidation ratio decreased from $\gamma = 17.2\%$ to 10.2%; and at $t = 3$ h, the matrix cracking density increased from $\lambda_m = 0.15$/mm at $\sigma_{mc} = 147$ MPa to $\lambda_m = 7.2$/mm at $\sigma_{sat} = 212$ MPa, and the interface oxidation ratio decreased from $\gamma = 40.7\%$ to 26.5%. At $T = 1073$ K and $t = 1$ h, the matrix cracking density increased from $\lambda_m = 0.15$/mm at $\sigma_{mc} = 132$ MPa to $\lambda_m = 7$/mm at $\sigma_{sat} = 198$ MPa, and the interface oxidation ratio decreased from $\gamma = 36.9\%$ to 22.4%; and at $t = 3$ h, the matrix cracking density increased from $\lambda_m = 0.09$/mm at $\sigma_{mc} = 132$ MPa to $\lambda_m = 4.8$/mm at $\sigma_{sat} = 198$ MPa, and the interface oxidation ratio decreased from $\gamma = 71\%$ to 50.1%.

Figure 3.12 shows the effect of interface debonding energy (i.e., $\Gamma_i = 0.3$, 0.5, and 0.7 J/m^2) on the time-dependent matrix cracking of SiC/SiC composites at $T = 873$, 973, and 1073 K for $t = 1$ and 3 h.

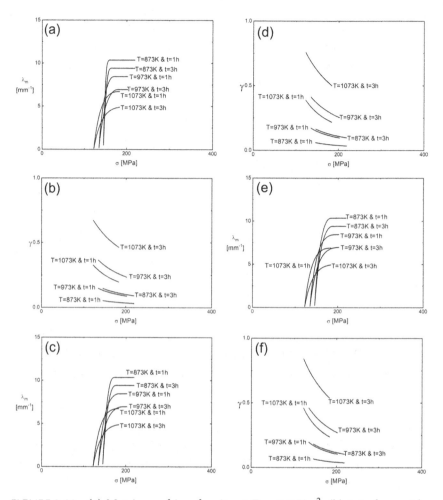

FIGURE 3.12 (a) Matrix cracking density at $\Gamma_i = 0.3$ J/m^2, (b) interface oxidation ratio at $\Gamma_i = 0.3$ J/m^2, (c) matrix cracking density at $\Gamma_i = 0.5$ J/m^2, (d) interface oxidation ratio $\Gamma_i = 0.5$ J/m^2, (e) matrix cracking density at $\Gamma_i = 0.7$ J/m^2, and (f) interface oxidation ratio at $\Gamma_i = 0.7$ J/m^2.

- When $\Gamma_i = 0.3$ J/m^2 at $T = 873$ K and $t = 1$ h, the matrix cracking density increased from $\lambda_m = 0.7$/mm at $\sigma_{mc} = 146$ MPa to $\lambda_m = 10.3$/mm at $\sigma_{sat} = 160$ MPa, and the interface oxidation ratio decreased from $\gamma = 5.2\%$ to 3.1%; and at $t = 3$ h, the matrix cracking density increased from $\lambda_m = 0.49$/mm at $\sigma_{mc} = 146$ MPa to $\lambda_m = 9.4$/mm at $\sigma_{sat} = 164$ MPa, and the interface oxidation ratio decreased from $\gamma = 14.7\%$ to 9.0%. At $T = 973$ K and $t = 1$ h, the matrix cracking density increased from $\lambda_m = 0.24$/mm at

σ_{mc} = 134 MPa to λ_m = 8.4/mm at σ_{sat} = 171 MPa, and the interface oxidation ratio decreased from y = 14.7% to 8.7%; and at t = 3 h, the matrix cracking density increased from λ_m = 0.16/mm at σ_{mc} = 134 MPa to λ_m = 6.9/mm at σ_{sat} = 179 MPa, and the interface oxidation ratio decreased from y = 36.5% to 23.3%. At T = 1073 K and t = 1 h, the matrix cracking density increased from λ_m = 0.15/mm at σ_{mc} = 122 MPa to λ_m = 6.7/mm at σ_{sat} = 183 MPa, and the interface oxidation ratio decreased from y = 32.6% to 19.6%; and at t = 3 h, the matrix cracking density increased from λ_m = 0.09/mm at σ_{mc} = 122 MPa to λ_m = 4.8/mm at σ_{sat} = 183 MPa, and the interface oxidation ratio decreased from y = 67.3% to 46.3%.

- When Γ_i = 0.5 J/m^2 at T = 873 K and t = 1 h, the matrix cracking density increased from λ_m = 0.3/mm at σ_{mc} = 146 MPa to λ_m = 10.3/mm at σ_{sat} = 173 MPa, and the interface oxidation ratio decreased from y = 5.9% to 3.3%; and at t = 3 h, the matrix cracking density increased from λ_m = 0.29/mm at σ_{mc} = 146 MPa to λ_m = 9.4/mm at σ_{sat} = 176 MPa, and the interface oxidation ratio decreased from y = 16.5% to 9.6%. At T = 973 K and t = 1 h, the matrix cracking density increased from λ_m = 0.2/mm at σ_{mc} = 134 MPa to λ_m = 8.4/mm at σ_{sat} = 184 MPa, and the interface oxidation ratio decreased from y = 17.1% to 9.5%; and at t = 3 h, the matrix cracking density increased from λ_m = 0.15/mm at σ_{mc} = 134 MPa to λ_m = 6.9/mm at σ_{sat} = 192 MPa, and the interface oxidation ratio decreased from y = 41.2% to 25.2%. At T = 1073 K and t = 1 h, the matrix cracking density increased from λ_m = 0.15/mm at σ_{mc} = 122 MPa to λ_m = 6.7/mm at σ_{sat} = 183 MPa, and the interface oxidation ratio decreased from y = 38.9% to 21.7%; and at t = 3 h, the matrix cracking density increased from λ_m = 0.09/mm at σ_{mc} = 122 MPa to λ_m = 4.8/mm at σ_{sat} = 183 MPa, and the interface oxidation ratio decreased from y = 75.6% to 50%.

- When Γ_i = 0.7 J/m^2 at T = 873 K and t = 1 h, the matrix cracking density increased from λ_m = 0.3/mm at σ_{mc} = 146 MPa to λ_m = 10.3/mm at σ_{sat} = 183 MPa, and the interface oxidation ratio decreased from y = 6.7% to 3.6%; and at t = 3 h, the matrix cracking density increased from λ_m = 0.24/mm at σ_{mc} = 146 MPa to λ_m = 9.4/mm at σ_{sat} = 186 MPa, and the interface oxidation ratio decreased from $\gamma = 18.2\%$ to 10.2%. At $T = 973$ K and $t = 1$ h, the matrix cracking density increased from $\lambda_m = 0.2$/mm at $\sigma_{mc} = 134$ MPa to $\lambda_m = 8.4$/mm at $\sigma_{sat} = 202$ MPa, and the interface oxidation ratio decreased from $\gamma = 19.7\%$ to 10.3%; and at $t = 3$ h, the matrix cracking density increased from $\lambda_m = 0.15$/mm at $\sigma_{mc} = 134$ MPa to $\lambda_m = 6.9$/mm at $\sigma_{sat} = 202$ MPa, and the interface oxidation ratio decreased from $\gamma = 46\%$ to 27%. At $T = 1073$ K and $t = 1$ h, the matrix cracking density increased from $\lambda_m = 0.16$/mm at $\sigma_{mc} = 122$ MPa to $\lambda_m = 6.8$/mm at $\sigma_{sat} = 183$ MPa, and the interface oxidation ratio decreased from $\gamma = 46\%$ to 23.8%; and at $t = 3$ h, the matrix cracking density increased from $\lambda_m = 0.09$/mm at $\sigma_{mc} = 122$ MPa to $\lambda_m = 4.9$/mm at $\sigma_{sat} = 183$ MPa, and the interface oxidation ratio decreased from $\gamma = 84.1\%$ to 53.6%.

Effect of the Matrix Property on Time-Dependent Multiple Matrix Cracking Evolution

Figure 3.13 shows the effect of matrix fracture energy (i.e., $\Gamma_m = 20, 25,$ and 30 J/m^2) on the time-dependent matrix cracking of SiC/SiC composites at $T = 873, 973,$ and 1073 K for $t = 1$ and 3 h.

- When $\Gamma_m = 20$ J/m^2 at $T = 873$ K and $t = 1$ h, the matrix cracking density increased from $\lambda_m = 0.3$/mm at $\sigma_{mc} = 162$ MPa to $\lambda_m = 8.4$/mm at $\sigma_{sat} = 195$ MPa, and the interface oxidation ratio decreased from $\gamma = 4.8\%$ to 2.8%; and at $t = 3$ h, the matrix cracking density increased from $\lambda_m = 0.24$/mm at $\sigma_{cr} = 162$ MPa to $\lambda_m = 7.8$/mm at $\sigma_{sat} = 199$ MPa, and the interface oxidation ratio decreased from $\gamma = 13.6\%$ to 8.2%. At $T = 973$ K and $t = 1$ h, the matrix cracking density increased from $\lambda_m = 0.18$/mm at $\sigma_{mc} = 149$ MPa to $\lambda_m = 7.1$/mm at $\sigma_{sat} = 204$ MPa, and the interface oxidation ratio decreased from $\gamma = 13.7\%$ to 8%; and at $t = 3$ h, the matrix cracking density increased from $\lambda_m = 0.13$/mm at $\sigma_{mc} = 149$ MPa to $\lambda_m = 6$/mm at $\sigma_{sat} = 212$ MPa, and the interface oxidation ratio decreased from $\gamma = 34.4\%$ to 21.6%. At $T = 1073$ K and $t = 1$ h, the matrix cracking density increased from $\lambda_m = 0.13$/mm at $\sigma_{mc} = 135$ MPa to $\lambda_m = 5.8$/mm at $\sigma_{sat} = 203$ MPa, and the interface oxidation ratio decreased from $\gamma = 30.9\%$ to 18.2%; and at $t = 3$ h, the matrix cracking density increased from $\lambda_m = 0.08$/mm at $\sigma_{cr} = 135$ MPa to $\lambda_m = 4.3$/mm at $\sigma_{sat} = 203$ MPa, and the interface oxidation ratio decreased from $\gamma = 64.7\%$ to 43.6%.

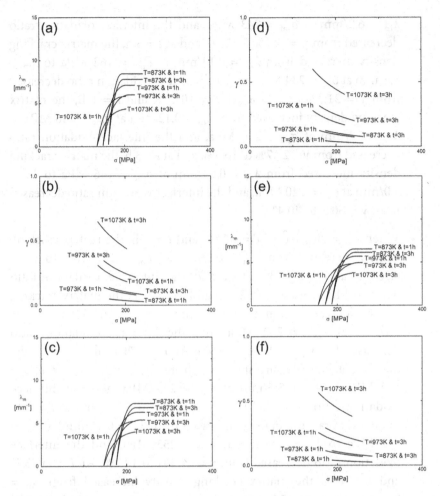

FIGURE 3.13 (a) Matrix cracking density at $\Gamma_m = 20$ J/m^2, (b) interface oxidation ratio at $\Gamma_m = 20$ J/m^2, (c) matrix cracking density at $\Gamma_m = 25$ J/m^2, (d) interface oxidation ratio at $\Gamma_m = 25$ J/m^2, (e) matrix cracking density at $\Gamma_m = 30$ J/m^2, and (f) interface oxidation ratio at $\Gamma_m = 30$ J/m^2.

- When $\Gamma_m = 25$ J/m^2 at $T = 873$ K and $t = 1$ h, the matrix cracking density increased from $\lambda_m = 0.23$/mm at $\sigma_{mc} = 175$ MPa to $\lambda_m = 7.3$/mm at $\sigma_{sat} = 218$ MPa, and the interface oxidation ratio decreased from $\gamma = 4.3\%$ to 2.6%; and at $t = 3$ h, the matrix cracking density increased from $\lambda_m = 0.2$/mm at $\sigma_{mc} = 175$ MPa to $\lambda_m = 6.8$/mm at $\sigma_{sat} = 222$ MPa, and the interface oxidation ratio decreased from $\gamma = 12.2\%$ to 7.5%. At $T = 973$ K and $t = 1$ h, the matrix cracking density increased from $\lambda_m = 0.15$/mm at $\sigma_{mc} = 162$ MPa to

λ_m = 6.2/mm at σ_{sat} = 243 MPa, and the interface oxidation ratio decreased from y = 12.3% to 7.3%; and at t = 3 h, the matrix cracking density increased from λ_m = 0.12/mm at σ_{mc} = 162 MPa to λ_m = 5.4/mm at σ_{sat} = 234 MPa, and the interface oxidation ratio decreased from y = 31.3% to 19.8%. At T = 1073 K and t = 1 h, the matrix cracking density increased from λ_m = 0.12/mm at σ_{mc} = 146 MPa to λ_m = 5.2/mm at σ_{sat} = 220 MPa, and the interface oxidation ratio decreased from y = 27.7% to 16.5%; and at t = 3 h, the matrix cracking density increased from λ_m = 0.08/mm at σ_{mc} = 146 MPa to λ_m = 4.0/mm at σ_{sat} = 220 MPa, and the interface oxidation ratio decreased from y = 60% to 40.4%.

- When Γ_m = 30 J/m^2 at T = 873 K and t = 1 h, the matrix cracking density increased from λ_m = 0.2/mm at σ_{mc} = 187 MPa to λ_m = 6.5/mm at σ_{sat} = 239 MPa, and the interface oxidation ratio decreased from y = 4% to 2.4%; and at t = 3 h, the matrix cracking density increased from λ_m = 0.17/mm at σ_{mc} = 167 MPa to λ_m = 6.1/mm at σ_{sat} = 242 MPa, and the interface oxidation ratio decreased from y = 11.3% to 6.9%. At T = 973 K and t = 1 h, the matrix cracking density increased from λ_m = 0.14/mm at σ_{mc} = 172 MPa to λ_m = 5.6/mm at σ_{sat} = 243 MPa, and the interface oxidation ratio decreased from y = 11.2% to 6.7%; and at t = 3 h, the matrix cracking density increased from λ_m = 0.11/mm at σ_{mc} = 172 MPa to λ_m = 4.9/mm at σ_{sat} = 253 MPa, and the interface oxidation ratio decreased from y = 29.1% to 18.4%. At T = 1073 K and t = 1 h, the matrix cracking density increased from λ_m = 0.11/mm at σ_{mc} = 156 MPa to λ_m = 4.8/mm at σ_{sat} = 234 MPa, and the interface oxidation ratio decreased from y = 25.4% to 15.3%; and at t = 3 h, the matrix cracking density increased from λ_m = 0.07/mm at σ_{mc} = 156 MPa to λ_m = 3.7/mm at σ_{sat} = 234 MPa, and the interface oxidation ratio decreased from y = 56.3% to 37.9%.

SUMMARY AND CONCLUSIONS

In this chapter, the effects of temperature and time on the matrix multiple cracking in CMCs were investigated. The CMSE criterion was adopted to determine the matrix cracking density as a function of applied stress, temperature, and duration at high temperatures. Experimental matrix cracking density of SiC/SiC composites corresponding to different temperatures and times was predicted. Effects of fiber volume fraction, interface properties, and matrix properties on the temperature/time-dependent matrix multiple cracking of SiC/SiC composites were also discussed.

REFERENCES

1. Li LB. Modeling matrix multi-fracture in SiC/SiC ceramic-matrix composites at elevated temperatures. *J. Aust. Ceram. Soc.* 2019; 55:1115–1126.
2. Li LB. Time-dependent matrix multi-fracture of SiC/SiC ceramic-matrix composites considering interface oxidation. *Ceramics-Silikaty* 2019; 63:131–148.
3. Li LB. Modeling first matrix cracking stress of fiber-reinforced ceramic-matrix composites considering fiber fracture. *Theor. Appl. Fract. Mech.* 2017; 92:24–32.
4. Choi SR, Gyekenyesi JP. Load-rate dependency of ultimate tensile strength in ceramic matrix composites at elevated temperatures. *Int. J. Fatigue* 2005; 27:503–510.
5. Liu S, Zhang L, Yin X, Liu Y, Cheng L. Proportional limit stress and residual thermal stress of 3D SiC/SiC composite. *J. Mater. Sci. Technol.* 2014; 30:959–964.
6. Sevener KM, Tracy JM, Chen Z, Kiser JD, Daly S. Crack opening behavior in ceramic matrix composites. *J. Am. Ceram. Soc.* 2017; 100:4734–4747.
7. Parthasarathy TA, Cox B, Surde O, Przybyla C, Cinibulk MK. Modeling environmentally induced property degradation of SiC/BN/SiC ceramic matrix composites. *J. Am. Ceram. Soc.* 2018; 101:973–997.
8. Sun Y, Singh RN. The generation of multiple matrix cracking and fiber-matrix interfacial debonding in a glass composite. *Acta Mater.* 1998; 46:1657–1667.
9. Cheng T, Qiao R, Xia Y. A Monte Carlo simulation of damage and failure process with crack saturation for unidirectional fiber reinforced ceramic composites. *Compos. Sci. Technol.* 2004; 64:2251–2260.
10. Morscher GN, Hee MY, DiCarlo JA. Matrix cracking in 3D orthogonal melt-infiltration SiC/SiC composite with various Z-fiber types. *J. Am. Ceram. Soc.* 2005; 88:146–153.
11. Rajan VP, Zok FW. Matrix cracking of fiber-reinforced ceramic composites in shear. *J. Mech. Phys. Solids* 2014; 73:3–21.
12. Gowayed Y, Ojard G, Santhosh U, Jefferson G. Modeling of crack density in ceramic matrix composites. *J. Compos. Mater.* 2015; 49:2285–2294.
13. Li L. Modeling for monotonic and cyclic tensile stress-strain behavior of 2D and 2.5D woven C/SiC ceramic-matrix composites. *Mech. Compos. Mater.* 2018; 54:165–178.
14. Li L. Micromechanical modeling for tensile behavior of carbon fiber-reinforced ceramic-matrix composites. *Appl. Compos. Mater.* 2015; 22:773–790.

15. Li L, Song Y, Sun Y. Modeling tensile behavior of cross-ply C/SiC ceramic-matrix composites. *Mech. Compos. Mater.* 2015; 51:358–376.

16. Li L, Song Y, Sun Y. Modeling tensile behavior of unidirectional C/SiC ceramic matrix composites. *Mech. Compos. Mater.* 2014; 49:659–672.

17. Guo S, Kagawa Y. Tensile fracture behavior of continuous SiC fiber-reinforced SiC matrix composites at elevated temperatures and correlation to in situ constituent properties. *J. Eur. Ceram. Soc.* 2002; 22:2349–2356.

18. Chen Z, Fang G, Xie J, Liang J. Experimental study of high-temperature tensile mechanical properties of 3D needled C/C-SiC composites. *Mater. Sci. Eng. A* 2016; 654:271–277.

19. Guo S, Kagawa Y. Effect of matrix modification on tensile mechanical behavior of Tyranno Si-Ti-C-O fiber-reinforced SiC matrix mini-composite at room and elevated temperatures. *J. Euro. Ceram. Soc.* 2004; 24:3261–3269.

20. Almansour A, Maillet E, Ramasamy R, Morscer GN. Effect of fiber content on single tow SiC minicomposite mechanical and damage properties using acoustic emission. *J. Eur. Ceram. Soc.* 2015; 35:3389–3399.

21. Morscher GN, Singh M, Kiser JD, Freedman M, Bhatt R. Modeling stress-dependent matrix cracking and stress-strain behavior in 2D woven SiC fiber reinforced CVI SiC composites. *Compos. Sci. Technol.* 2007; 67:1009–1017.

22. Aveston J, Cooper GA, Kelly A. In: The Properties of Fiber Composites, Conference on Proceedings. National Physical Laboratory. Guildford: IPC Science and Technology Press, 1971; 15–26.

23. Budiansky B, Hutchinson J W, Evans AG. Matrix fracture in fiber-reinforced ceramics. *J. Mech. Phys. Solids* 1986; 34:167–189.

24. Guo S, Kagawa Y. Temperature dependence of in situ constituent properties of polymer-infiltration-pyrolysis-processed Nicalon™ SiC fiber-reinforced SiC matrix composite. *J. Mater. Res.* 2000; 15:951–960.

25. Filipuzzi L, Camus G, Thebault J, Naslain R. Oxidation mechanisms and kinetics of 1D-SiC/SiC composite materials: I, An experimental approach. *J. Am. Ceram. Soc.* 1994; 77:459–466.

26. Lamouroux F, Jouin JM, Naslain R. Kinetics and mechanics of oxidation of 2D woven C/SiC composites: II, Theoretical approach. *J. Am. Ceram. Soc.* 1994; 77:2058–2068.

27. Verrilli MJ, Opila EJ, Calomino A, Kiser JD. Effect of environment on the stress-rupture behavior of a carbon-fiber-reinforced silicon carbide ceramic matrix composite. *J. Am. Ceram. Soc.* 2004; 87:1536–1542.

28. Halbig MC, McGuffin-Cawley JD, Eckel AD, Brewer DN. Oxidation kinetics and stress effects for the oxidation of continuous carbon fibers within a microcracked C/SiC ceramic matrix composite. *J. Am. Ceram. Soc.* 2008; 91:519–526.

29. Li LB. Modeling matrix cracking of fiber-reinforced ceramic-matrix composites under oxidation environment at elevated temperature. *Theor. Appl. Fract. Mech.* 2017; 87:110–119.

30. Smith CE, Morscher GN, Xia ZH. Monitoring damage accumulation in ceramic matrix composites using electrical resistivity. *Scr. Mater.* 2008; 59:463–466.
31. Simon C, Rebillat F, Herb V, Camus G. Monitoring damage evolution of $SiC_f/[Si-B-C]_m$ composites using electrical resistivity: Crack density-based electromechanical modeling. *Acta Mater.* 2017; 124:579–587.
32. Gowayed Y, Ojard G, Santhosh U, Jefferson G. Modeling of crack density in ceramic matrix composites. *J. Compos. Mater.* 2015; 49:2285–2294.
33. Parthasarathy TA, Cox B, Surde O, Przybyla C, Cinibulk MK. Modeling environmentally induced property degradation of SiC/BN/SiC ceramic matrix composites. *J. Am. Ceram. Soc.* 2018; 101:973–997.
34. Zhang S, Gao X, Chen J, Dong H, Song Y. Strength model of the matrix element in SiC/SiC composites. *Mater. Design* 2016; 101:66–71.

CHAPTER 4

High-Temperature Crack Opening Behavior in Ceramic-Matrix Composites

INTRODUCTION

The excellent non-brittle fracture behavior of ceramic-matrix composites (CMCs) is mainly due to the optimization design of fiber/matrix inter-phase to realize matrix crack deflection on the fiber surface, fiber bridging at the crack plane, fiber fracture, and relative slip friction between the fiber and the matrix to consume strain energy [1–3]. The matrix fracture strain of CMCs is much lower than that of the fiber [4,5]. In the stress environment, matrix cracks occur first [6]. The appearance of cracks increases the oxygen diffusion channel and accelerates the oxidation failure of CMCs [7–10]. Forna-Kreutzer et al. [11] observed the interface cracks, inclined cracks, and crack opening behavior in Nextel^{TM} 720/Alumina oxide-oxide composites at room temperature and 1050°C using X-ray computed micro-tomography (XCT) and digital volume correlation (DVC). Jordan et al. [12] characterized the initiation, distribution, and development of matrix cracks and crack opening displacement (COD) under tensile and fatigue loading in a SiC/SiC composite using the XCT technique. Relationship between saturation of matrix cracks and fiber/matrix interfacial mechanics was established. Chen et al. [13] detected the crack networks in SiC/SiC tubes and estimated the crack opening, orientation, and surface area of the detected

DOI: 10.1201/9781032638508-4

cracks by combining the DVC and XCT techniques. Due to the thermal misfit and matrix internal defects, brittle fracture occurs in the SiC matrix with increasing applied stress. Opening of these cracks will control the oxidation atmosphere (i.e., oxygen and/or water vapor) to ingress into the composites, oxidizing the fiber and the interphase over a period of time or cycle and reducing the strength and durability of CMCs [7,14,15]. Chateau et al. [16] performed *in situ* tensile experiments on SiC/SiC minicomposites and obtained the COD curves under tensile loading. The length of matrix crack affects the crack opening behavior. Sevener et al. [17] performed an experimental investigation on the opening of matrix cracking and interface in 2D plain-woven SiC/SiC composites. Cracks in transverse and longitudinal yarns appear in different COD behavior. With increasing applied stress, matrix cracking density increases and gradually approaches saturation [18,19].

The objective of this chapter is to analyze the stress-dependent, cyclic-dependent, and time-dependent crack opening behavior in SiC/SiC composites. Multiple damage mechanisms were considered. At the matrix cracking plane, the COD and related fiber and matrix axial displacements were derived. Experimental CODs of SiC/SiC composite were predicted. Effects of composite's constituent properties, stress level, and temperature on crack opening behavior in SiC/SiC composite were also analyzed.

STRESS-DEPENDENT CRACK OPENING BEHAVIOR IN CMCS

In this section, a micromechanical approach was developed to predict the CODs of fiber-reinforced CMCs under tensile loading. The length of matrix cracks was divided into long, medium, and short. Experimental CODs of SiC/SiC composites were predicted. Effects of composite's constituent properties, matrix crack length, and interface shear stress on CODs, crack opening stress (COS), interface debonding ratio (IDR), interface complete debonding stress (ICDS), and axial fiber and matrix displacements at the crack plane were also analyzed.

Micromechanical Model

For the long matrix crack length, the fiber and matrix axial stresses are different in the interface slip region ($x \in [0, l_\mathrm{d}]$) and interface bonding region $\left(x \in \left[l_\mathrm{d}, \frac{l_\mathrm{c}}{2}\right]\right)$. The fiber and matrix axial displacements, $w_\mathrm{f}(x)$ and $w_\mathrm{m}(x)$, are

$$w_f(x) = \int_x^{l_c/2} \frac{\sigma_f}{E_f} dx$$

$$= \frac{\sigma}{V_f E_f}(l_d - x) + \frac{\sigma_{fo}}{E_f}\left(\frac{l_c}{2} - l_d\right) - \frac{\tau_i}{r_f E_f}(l_d^2 - x^2) \tag{4.1}$$

$$- \frac{r_f}{\rho E_f}\left(\frac{V_m}{V_f}\sigma_{mo} - 2\frac{l_d}{r_f}\tau_i\right)\left[\exp\left(-\rho\frac{l_c/2 - l_d}{r_f}\right) - 1\right]$$

$$w_m(x) = \int_x^{l_c/2} \frac{\sigma_m}{E_m} dx$$

$$= \frac{V_f \tau_i}{r_f V_m E_m}(l_d^2 - x^2) + \frac{\sigma_{mo}}{E_m}\left(\frac{l_c}{2} - l_d\right) \tag{4.2}$$

$$+ \frac{r_f}{\rho E_m}\left(\sigma_{mo} - 2\tau_i\frac{V_f}{V_m}\frac{l_d}{r_f}\right)\left[\exp\left(-\rho\frac{l_c/2 - l_d}{r_f}\right) - 1\right]$$

The relative displacement between the fiber and the matrix, $u(x)$, is

$$u(x) = w_f(x) - w_m(x)$$

$$= \frac{\sigma}{V_f E_f}(l_d - x) + \left(\frac{\sigma_{fo}}{E_f} - \frac{\sigma_{mo}}{E_m}\right)\left(\frac{l_c}{2} - l_d\right) - \frac{E_c \tau_i}{r_f V_m E_f E_m}(l_d^2 - x^2) \tag{4.3}$$

$$- \frac{r_f E_c}{\rho V_f E_f E_m}\left(\sigma_{mo} - 2\frac{V_f}{V_m}\frac{l_d}{r_f}\tau_i\right)\left[\exp\left(-\rho\frac{l_c/2 - l_d}{r_f}\right) - 1\right]$$

The interface debonding length, l_d, can be determined by the fracture mechanic approach, which is given by

$$\Gamma_i = \frac{F}{4\pi r_f}\frac{\partial w_f(x = 0)}{\partial l_d} - \frac{1}{2}\int_0^{l_d} \tau_i\frac{\partial u(x)}{\partial l_d}dx \tag{4.4}$$

where Γ_i is the interface debonding energy.

Substituting Eqs. (4.1) and (4.3) into Eq. (4.4), the interface debonding length is

$$l_d = \frac{r_f}{2}\left(\frac{V_m E_m \sigma}{V_f E_c \tau_i} - \frac{1}{\rho}\right) - \sqrt{\left(\frac{r_f}{2\rho}\right)^2 + \frac{r_f V_m E_m E_f}{E_c \tau_i^2}\Gamma_i} \tag{4.5}$$

At the matrix crack plane, the COD is

$$\text{COD} = 2u_{cod} = \frac{\sigma}{V_f E_f}l_c\eta + \left(\frac{\sigma_{fo}}{E_f} - \frac{\sigma_{mo}}{E_m}\right)l_c(1 - \eta) - \frac{E_c \tau_i}{2r_f V_m E_f E_m}(l_c\eta)^2$$

$$- \frac{2r_f E_c}{\rho V_f E_f E_m}\left(\sigma_{mo} - \frac{V_f}{V_m}\frac{l_c}{r_f}\eta\tau_i\right)\left[\exp\left(-\rho\frac{l_c}{2r_f}(1 - \eta)\right) - 1\right]$$

$$\tag{4.6}$$

where η is the IDR.

For the medium and short matrix crack length, the interface debonding length occupies the entire matrix fragmentation length. The fiber and matrix axial displacements, $w_f(x)$ and $w_m(x)$, are as follows:

$$w_f(x) = \int_x^{l_c/2} \frac{\sigma_f}{E_f} dx = \frac{\sigma}{V_f E_f}\left(\frac{l_c}{2} - x\right) - \frac{\tau_i}{r_f E_f}\left[\left(\frac{l_c}{2}\right)^2 - x^2\right] \quad (4.7)$$

$$w_m(x) = \int_x^{l_c/2} \frac{\sigma_m}{E_m} dx = \frac{V_f \tau_i}{4 r_f V_m E_m}(l_c^2 - 4x^2) \quad (4.8)$$

The relative displacement between the fiber and the matrix, $u(x)$, is

$$u(x) = w_f(x) - w_m(x) = \frac{\sigma}{2 V_f E_f}(l_c - 2x) - \frac{E_c \tau_i}{4 r_f V_m E_f E_m}(l_c^2 - 4x^2) \quad (4.9)$$

At the matrix crack plane, COD can be determined by

$$COD = 2u_{cod} = \frac{\sigma}{V_f E_f} l_c - \frac{E_c \tau_i}{2 r_f V_m E_f E_m} l_c^2 \quad (4.10)$$

Experimental Comparisons

Chateau et al. [16] measured the COD of Hi-Nicalon^{TM} type S SiC/SiC minicomposites through *in situ* tensile tests under a scanning electron microscope at room temperature.

Figure 4.1 shows the experimental and predicted COD (u_{cod}), COS (σ_{cos}), IDR (η), ICDS (σ_{icds}), and fiber and matrix axial displacements (w_f and w_m) for different matrix crack lengths.

- For matrix crack #1, the COS was approximately $\sigma_{cos} = 592$ MPa; the experimental u_{cod} increased from $u_{cod} = 0.18$ to 0.85 μm while increasing applied stress from $\sigma = 627$ to 774 MPa, and the predicted u_{cod} increased from $u_{cod} = 0.03$ to 0.88 μm while increasing applied stress from $\sigma = 592$ to 802 MPa; the IDR increased from $\eta = 0.06$ to 0.73 while increasing applied stress from $\sigma = 592$ to 802 MPa; the fiber axial displacement at the crack plane increased from $w_f = 0.67$ to 1.22 μm while increasing applied stress from $\sigma = 592$ to 802 MPa; and the matrix axial displacement at the crack

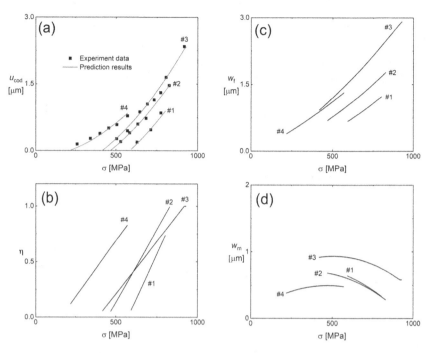

FIGURE 4.1   (a) u_{cod}, (b) η, (c) w_f, and (d) w_m of SiC/SiC minicomposites at room temperature.

plane decreased from w_m = 0.64 to 0.33 μm while increasing applied stress from σ = 592 to 802 MPa.

- For matrix crack #2, the COS was approximately σ_{cos} = 469 MPa; the experimental u_{cod} increased from u_{cod} = 0.19 to 1.46 μm while increasing applied stress from σ = 528 to 824 MPa, and the predicted u_{cod} increased from u_{cod} = 0.007 to 1.48 μm while increasing applied stress from σ = 469 to 829 MPa; the IDR increased from η = 0.052 to 0.99 while increasing applied stress from σ = 469 to 829 MPa; the fiber axial displacement at the crack plane increased from w_f = 0.69 to 1.77 μm while increasing applied stress from σ = 469 to 829 MPa; and the matrix axial displacement at the crack plane decreased from w_m = 0.68 to 0.28 μm while increasing applied stress from σ = 469 to 829 MPa.

- For matrix crack #3, the COS was approximately σ_{cos} = 419 MPa; the experimental u_{cod} increased from u_{cod} = 0.26 to 2.34 μm while increasing applied stress from σ = 504 to 917 MPa, and the predicted u_{cod} increased from u_{cod} = 0.006 to 2.33 μm while increasing applied stress from $\sigma = 419$ to 929 MPa; the IDR increased from $\eta = 0.05$ to 1.0 while increasing applied stress from $\sigma = 419$ to 919 MPa; the fiber axial displacement at the crack plane increased from $w_f = 0.92$ to 2.9 μm while increasing applied stress from $\sigma = 419$ to 929 MPa; and the matrix axial displacement at the crack plane decreased from $w_m = 0.92$ to 0.58 μm while increasing applied stress from $\sigma = 419$ to 929 MPa.

- For matrix crack #4, the COS was approximately $\sigma_{cos} = 219$ MPa; the experimental u_{cod} increased from $u_{cod} = 0.14$ to 0.78 μm while increasing applied stress from $\sigma = 260$ to 569 MPa, and the predicted u_{cod} increased from $u_{cod} = 0.008$ to 0.83 μm while increasing applied stress from $\sigma = 219$ to 569 MPa; the IDR increased from $\eta = 0.12$ to 0.83 while increasing applied stress from $\sigma = 219$ to 569 MPa; the fiber axial displacement at the crack plane increased from $w_f = 0.39$ to 1.32 μm while increasing applied stress from $\sigma = 219$ to 569 MPa; and the matrix axial displacement at the crack plane increased from $w_m = 0.38$ to 0.496 μm while increasing applied stress from $\sigma = 219$ to 469 MPa, and then decreased to $w_m = 0.48$ μm at $\sigma = 569$ MPa.

Discussions

In this section, the effects of fiber volume fraction, fiber, matrix, and interface properties on stress-dependent crack opening behavior of SiC/SiC minicomposites were analyzed.

Effect of Fiber Volume on Stress-Dependent CODs

Figure 4.2 shows the effect of fiber volume fraction (i.e., $V_f = 30\%$, 40%, and 50%) on the matrix crack opening behavior of SiC/SiC composites. When $V_f = 30\%$, 40%, and 50%, the COD increased with applied stress and can be divided into two regions: partial interface debonding (PID) region and complete interface debonding (CID) region. The increased rate of COD for PID is much higher than that for CID. For the damage state of PID, the fiber axial displacement at the matrix cracking plane increased with the applied stress; however, the matrix axial displacement at the matrix cracking plane decreased with the applied stress. For the damage state of the CID, the fiber axial displacement at the matrix cracking plane still increased with the applied stress; however, the increased rate in fiber axial displacement for CID was much lower than

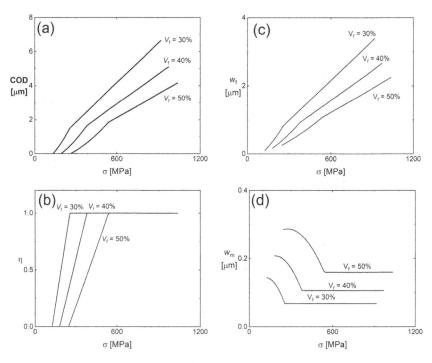

FIGURE 4.2   Effect of fiber volume fraction on (a) COD, (b) η, (c) w_f, and (d) w_m.

that for PID, and the matrix axial displacement at the matrix cracking plane remained constant. When the fiber volume fraction increased from $V_f = 30\%$ to 50%, the COS increased from $\sigma_{cos} = 136$ to 266 MPa, the ICDS increased from $\sigma_{icds} = 256$ to 546 MPa, and the COD decreased at the same applied stress, due to the decrease in the IDR.

Effect of Fiber Properties on Stress-Dependent CODs

Figure 4.3 shows the effect of fiber radius (i.e., $r_f = 5$, 6, and 7 μm) on the matrix crack opening behavior of SiC/SiC composites. When $r_f = 5$, 6, and 7 μm, the COD and the IDR both increased with the increase in fiber radius. For PID, the fiber axial displacement at the matrix cracking plane increased with an increase in applied stress; however, the matrix axial displacement at the matrix cracking plane decreased with an increase in applied stress. For CID, the fiber axial displacement at the matrix cracking plane still increased with the applied stress; however, the increased rate for CID was lower than that for PID, and the matrix axial displacement at the matrix cracking plane remained constant. The fiber axial displacement at the matrix cracking plane increased with fiber

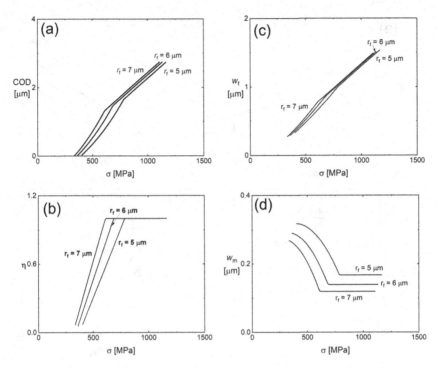

FIGURE 4.3 Effect of fiber radius on (a) COD, (b) η, (c) w_f, and (d) w_m.

radius; however, the matrix axial displacement decreased with fiber radius. When fiber radius was increased from $r_f = 5$ to 7 μm, the COS decreased from $\sigma_{cos} = 403$ to 336 MPa and the ICDS also decreased from $\sigma_{icds} = 793$ to 615 MPa.

Figure 4.4 shows the effect of fiber elastic modulus (i.e., $E_f = 250$, 300, and 350 GPa) on the matrix crack opening behavior of SiC/SiC composites. When $E_f = 250$, 300, and 350 GPa, the COD and IDR both decreased with an increase in fiber elastic modulus. For PID, the fiber axial displacement at the matrix cracking plane increased with the increase in applied stress; however, the matrix axial displacement at the cracking plane decreased with the increase in applied stress. For CID, the fiber axial displacement at the cracking plane still increased with the increase in applied stress; however, the increased rate of fiber axial displacement for CID was much lower than that for PID, and the matrix axial displacement at the crack plane remained constant. The fiber axial displacement at the cracking plane decreased with the increase in fiber elastic modulus; however, the matrix axial displacement at the crack plane increased with the increase in fiber elastic modulus for PID, and

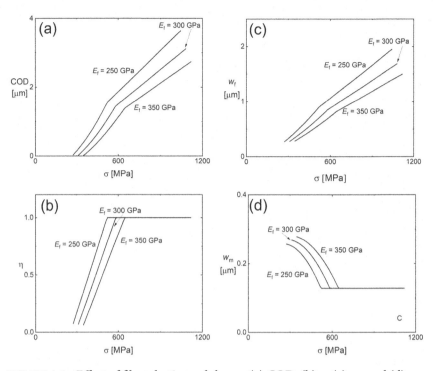

FIGURE 4.4   Effect of fiber elastic modulus on (a) COD, (b) η, (c) w_f, and (d) w_m.

for CID, the matrix axial displacement at the crack plane remained constant. When the fiber elastic modulus increased from E_f = 250 to 350 GPa, the COS increased from σ_{cos} = 273 to 347 MPa, and the ICDS increased from σ_{icds} = 523 to 647 MPa.

Effect of Matrix Properties on Stress-Dependent CODs
Figure 4.5 shows the effect of matrix elastic modulus (i.e., E_m = 200, 300, and 400 GPa) on the matrix crack opening behavior of SiC/SiC composites. For matrix elastic modulus E_m = 200, 300, and 400 GPa, the COD and IDR both increased with the increase in matrix elastic modulus. For PID, the fiber axial displacement at the cracking plane increased with the increase in applied stress; however, the matrix axial displacement at the cracking plane decreased with applied stress. For CID, the fiber axial displacement at the cracking plane still increased with applied stress; however, the increased rate of fiber axial displacement for CID was much lower than that for PID, and the matrix axial displacement at the cracking plane remained constant. The fiber axial displacement at the cracking plane increased with the increase in matrix

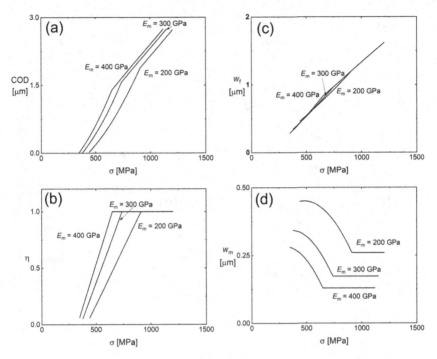

FIGURE 4.5   Effect of matrix elastic modulus on (a) COD, (b) η, (c) w_f, and (d) w_m.

elastic modulus for PID and remained the same for CID. The matrix axial displacement at the crack plane decreased with the increase in the matrix elastic modulus. When the matrix elastic modulus increased from $E_m = 200$ to 400 GPa, the COS decreased from $\sigma_{cos} = 441$ to 348 MPa, and the ICDS decreased from $\sigma_{icds} = 921$ to 658 MPa.

Effect of Interface Properties on Stress-Dependent CODs
Figure 4.6 shows the effect of interface shear stress (i.e., $\tau_i = 5, 10,$ and 15 MPa) on crack opening behavior of SiC/SiC composites. For the interface shear stress $\tau_i = 5$ MPa, the COD versus applied stress curve was divided into two regions corresponding to the PID region and CID region. For interface shear stress $\tau_i = 10$ and 15 MPa, the COD versus applied stress curve has only one region corresponding to the PID region. The fiber and matrix axial displacements at the matrix cracking plane both decreased with the increase in interface shear stress from $\tau_i = 5$ to 15 MPa. When the interface shear stress increased from $\tau_i = 5$ to 15 MPa, the COS increased from $\sigma_{cos} = 368$ to 431 MPa, and the ICDS also increased.

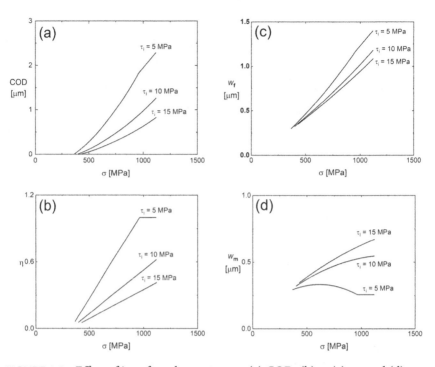

FIGURE 4.6   Effect of interface shear stress on (a) COD, (b) η, (c) w_f, and (d) w_m.

CYCLIC-DEPENDENT CRACK OPENING BEHAVIOR IN CMCS

In this section, the cyclic-dependent crack opening behavior in CMCs considering the effect of interface wear was investigated. Effects of composite's constituent properties, peak stress level, and damage state on crack opening behavior of CMCs were analyzed. Theoretical relationships between the COD, peak stress, cycle number, and composite's constitutive properties were established.

Micromechanical Model

Under cyclic loading, fiber axial displacement w_f and matrix axial displacement w_m were given by

$$
\begin{aligned}
w_f(x) &= \int_x^{l_c/2} \frac{\sigma_f(\sigma, N, x)}{E_f} dx \\
&= \frac{\sigma}{V_f E_f}\left(l_d(\sigma, N) - x\right) + \frac{\sigma_{fo}}{E_f}\left(\frac{l_c}{2} - l_d(\sigma, N)\right) - \frac{\tau_i(N)}{r_f E_f}\left(l_d^2(\sigma, N) - x^2\right) \\
&\quad - \frac{r_f}{\rho E_f}\left(\frac{V_m}{V_f}\sigma_{mo} - 2\frac{l_d(\sigma, N)}{r_f}\tau_i(N)\right)\left[\exp\left(-\rho\frac{l_c/2 - l_d(\sigma, N)}{r_f}\right) - 1\right]
\end{aligned}
$$

$$(4.11)$$

$$w_m(x) = \int_x^{l_c/2} \frac{\sigma_m(\sigma, N, x)}{E_m} dx$$

$$= \frac{V_f \tau_i(N)}{r_f V_m E_m} (l_d^2(\sigma, N) - x^2) + \frac{\sigma_{mo}}{E_m} \left(\frac{l_c}{2} - l_d(\sigma, N) \right)$$

$$+ \frac{r_f}{\rho E_m} \left(\sigma_{mo} - 2\tau_i(N) \frac{V_f}{V_m} \frac{l_d(\sigma, N)}{r_f} \right) \left[\exp\left(-\rho \frac{l_c/2 - l_d(\sigma, N)}{r_f} \right) - 1 \right]$$

(4.12)

Using Eqs. (4.11) and (4.12), the relative displacement $u(x)$ is

$$u(x) = w_f(x) - w_m(x)$$

$$= \frac{\sigma}{V_f E_f} (l_d(\sigma, N) - x) + \left(\frac{\sigma_{fo}}{E_f} - \frac{\sigma_{mo}}{E_m} \right) \left(\frac{l_c}{2} - l_d(\sigma, N) \right)$$

$$- \frac{E_c \tau_i(N)}{r_f V_m E_f E_m} (l_d^2(\sigma, N) - x^2)$$

$$- \frac{r_f E_c}{\rho V_f E_f E_m} \left(\sigma_{mo} - 2 \frac{V_f}{V_m} \frac{l_d(\sigma, N)}{r_f} \tau_i(N) \right) \left[\exp\left(-\rho \frac{l_c/2 - l_d(\sigma, N)}{r_f} \right) - 1 \right]$$

(4.13)

At the matrix cracking plane, the COD is

$$COD = 2u_{cod}$$

$$= \frac{\sigma}{V_f E_f} l_c \eta + \left(\frac{\sigma_{fo}}{E_f} - \frac{\sigma_{mo}}{E_m} \right) l_c (1 - \eta) - \frac{E_c \tau_i(N)}{2r_f V_m E_f E_m} (l_c \eta)^2$$

$$- \frac{2r_f E_c}{\rho V_f E_f E_m} \left(\sigma_{mo} - \frac{V_f}{V_m} \frac{l_c}{r_f} \eta \tau_i(N) \right) \left[\exp\left(-\rho \frac{l_c}{2r_f} (1 - \eta) \right) - 1 \right]$$

(4.14)

With the increase in applied cycles, the interface debonding length propagates along the fiber/matrix interface and approaches the matrix crack spacing at cycle number N_{CID}, i.e., $\eta (N = N_{CID}) = 1.0$. The relative displacement $u(x)$ is

$$u(x) = w_f(x) - w_m(x) = \frac{\sigma}{2V_f E_f} (l_c - 2x) - \frac{E_c \tau_i(N)}{4r_f V_m E_f E_m} (l_c^2 - 4x^2)$$

(4.15)

At the matrix cracking plane, the COD is

$$COD = 2u_{cod} = \frac{\sigma}{V_f E_f} l_c - \frac{E_c \tau_i(N)}{2r_f V_m E_f E_m} l_c^2$$

(4.16)

Experimental Comparisons

Figure 4.7 shows the experimental and analytical predicted u_{cod} and related damage parameters with applied stress of SiC/SiC crack #1 for different cycles.

When $N = 1$, u_{cod} increased nonlinearly with applied stress. The increasing rate was relatively low upon initial loading due to the short interface debonding range. However, the increasing rate of u_{cod} versus σ curve decreased at high stress for the CID condition. η increased linearly with applied tensile stress till CID. Due to the existence of the interface bonding region in the matrix crack spacing, fiber axial displacement increased nonlinearly with stress for the PID condition (i.e., $\eta < 1.0$). The increasing rate of fiber axial displacement versus σ curve increased till CID (i.e., $\eta = 1.0$), at which the increasing rate of fiber axial displacement versus σ curve remained constant. For the short interface debonding range condition, the matrix axial displacement increased with stress and approached peak value. With the increase in applied stress, the interface bonding range decreased and the load transfer capacity between the fiber

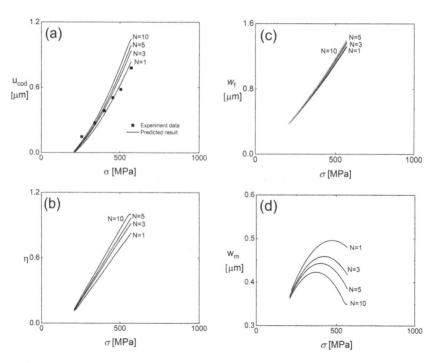

FIGURE 4.7 Experimental and predicted crack opening behavior of SiC/SiC crack #1 for $N = 1, 3, 5,$ and 10: (a) u_{cod}, (b) η, (c) w_f, and (d) w_m.

and the matrix decreased, leading to the decrease in matrix axial displacement with applied stress till CID (i.e., $\eta = 1$), at which the matrix axial displacement remained constant with increasing stress.

When the cycle number increased from $N = 1$ to 10, the value of u_{cod} increased at the same stress level, and the incremental range increased with stress level; when $N = 1$, 3, and 5, PID (i.e., $\eta < 1$) occurred with increasing stress, and when $N = 10$, the CID (i.e., $\eta = 1$) occurred at high stress.

- When $N = 1$, u_{cod} increased from $u_{cod} = 0.008$ to 0.837 μm with the increase in applied stress from $\sigma = 219$ to 569 MPa; η increased from $\eta = 0.12$ to 0.82; w_f increased from $w_f = 0.389$ to 1.361 μm; and w_m increased nonlinearly from $w_m = 0.38$ to 0.496 μm and then decreased to 0.479 μm.

- When $N = 3$, u_{cod} increased from $u_{cod} = 0.0012$ to 0.936 μm with the increase in applied stress from $\sigma = 209$ to 569 MPa; η increased from $\eta = 0.112$ to 0.918; w_f increased from $w_f = 0.369$ to 1.353 μm; and w_m increased nonlinearly from $w_m = 0.368$ to 0.458 μm and then decreased to 0.416 μm.

- When $N = 5$, u_{cod} increased from $u_{cod} = 0.0046$ to 0.987 μm with the increase in applied stress from $\sigma = 209$ to 569 MPa; η increased from $\eta = 0.117$ to 0.965; w_f increased from $w_f = 0.37$ to 1.37 μm; and w_m increased nonlinearly from $w_m = 0.366$ to 0.443 μm and then decreased to 0.384 μm.

- When $N = 10$, u_{cod} increased from $u_{cod} = 0.009$ to 1.043 μm with the increase in applied stress from $\sigma = 209$ to 569 MPa; η increased from $\eta = 0.126$ to 1.0; w_f increased from $w_f = 0.372$ to 1.393 μm; and w_m increased nonlinearly from $w_m = 0.362$ to 0.423 μm and then decreased to 0.349 μm.

Figure 4.8 shows the experimental and predicted u_{cod} and related damage parameters with applied stress of SiC/SiC crack #2 for different cycles. Compared with crack #1, CID (i.e., $\eta = 1.0$) occurred for different cycle numbers, and the applied stress corresponding to the CID condition decreased with the cycle number. Evolution of u_{cod}, w_f, and w_m versus σ curves for $N = 1$, 3, 5, and 10 can be divided into two regions, i.e., the nonlinearly increasing region for the PID condition ($\eta < 1.0$) and the linearly increasing region for the CID condition ($\eta = 1.0$).

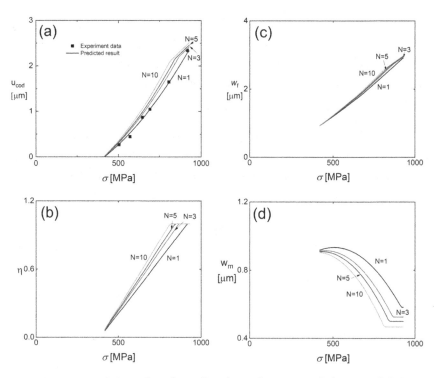

FIGURE 4.8 Experimental and predicted crack opening behavior of SiC/SiC crack #2 for $N = 1$, 3, 5, and 10: (a) u_{cod}; (b) η, (c) w_f, and (d) w_m.

- When $N = 1$, u_{cod} increased from $u_{cod} = 0.0058$ to 2.329 μm with applied stress increasing from $\sigma = 419$ to 929 MPa; η increased from $\eta = 0.056$ to 1.0; w_f increased from $w_f = 0.923$ to 2.911 μm; w_m increased nonlinearly from $w_m = 0.917$ to 0.933 μm, then decreased to 0.582 μm, and remained unchanged. The CID stress was $\sigma_{cid} = 918$ MPa, the corresponding $u_{cod} = 2.29$ μm, $w_f = 2.87$ μm, and $w_m = 0.58$ μm.

- When $N = 3$, u_{cod} increased from $u_{cod} = 0.015$ to 2.419 μm with applied stress increasing from $\sigma = 419$ to 929 MPa; η increased from $\eta = 0.062$ to 1.0; w_f increased from $w_f = 0.926$ to 2.945 μm; w_m increased nonlinearly from $w_m = 0.911$ to 0.914 μm, then decreased to 0.525 μm, and remained unchanged. The CID stress is $\sigma_{cid} = 868$ MPa, the corresponding $u_{cod} = 2.2$ μm, $w_f = 2.73$ μm, and $w_m = 0.52$ μm.

- When $N = 5$, u_{cod} increased from $u_{cod} = 0.019$ to 2.46 μm with applied stress increasing from $\sigma = 419$ to 929 MPa; η increased from

η = 0.066 to 1.0; w_f increased from w_f = 0.928 to 2.959 μm; w_m increased nonlinearly from w_m = 0.908 to 0.909 μm, then decreased to 0.499 μm, and remained unchanged. The CID stress was σ_{cid} = 848 MPa, the corresponding u_{cod} = 2.18 μm, w_f = 2.67 μm, and w_m = 0.49 μm.

- When N = 10, u_{cod} increased from u_{cod} = 0.027 to 2.51 μm with applied stress increasing from σ = 419 to 929 MPa; η increased from η = 0.07 to 1.0; and w_f increased from w_f = 0.931 to 2.979 μm; w_m decreased nonlinearly from w_m = 0.904 to 0.466 μm and remained unchanged. The CID stress was σ_{cid} = 818 MPa, the corresponding u_{cod} = 2.12 μm, w_f = 2.59 μm, and w_m = 0.46 μm.

Figure 4.9 shows experimental and predicted u_{cod} and related damage parameters. When N = 1, PID occurred, and with the increase in applied cycles to N = 3, 5, and 10, the CID occurred and the applied stress for CID decreased with the increase in applied cycles. When N = 1, u_{cod}, w_f,

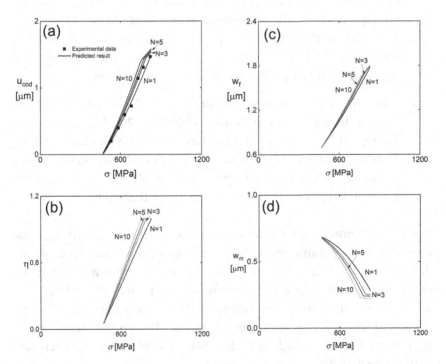

FIGURE 4.9 Experimental and predicted crack opening behavior of SiC/SiC crack #3 for N = 1, 3, 5, and 10: (a) u_{cod}, (b) η, (c) w_f, and (d) w_m.

and w_m increased nonlinearly with increasing stress; when $N = 3$, 5, and 10, u_{cod}, w_f, and w_m versus σ curves can be divided into two regions, i.e., the nonlinearly increasing region for PID (i.e., $\eta < 1.0$) and the linearly increasing region for CID (i.e., $\eta = 1.0$).

- When $N = 1$, u_{cod} increased from $u_{cod} = 0.0072$ to 1.486 μm with the increase in applied stress from $\sigma = 469$ to 829 MPa; η increased from $\eta = 0.052$ to 0.994; w_f increased from $w_f = 0.69$ to 1.769 μm; w_m decreased nonlinearly from $w_m = 0.683$ to 0.283 μm.

- When $N = 3$, u_{cod} increased from $u_{cod} = 0.014$ to 1.535 μm with the increase in applied stress from $\sigma = 469$ to 829 MPa; η increased from $\eta = 0.058$ to 1.0; w_f increased from $w_f = 0.692$ to 1.787 μm; w_m decreased nonlinearly from $w_m = 0.678$ to 0.252 μm and remained unchanged. The CID stress was $\sigma_{cid} = 799$ MPa, the corresponding $u_{cod} = 1.465$ μm, $w_f = 1.717$ μm, and $w_m = 0.252$ μm.

- When $N = 5$, u_{cod} increased from $u_{cod} = 0.017$ to 1.554 μm with the increase in applied stress from $\sigma = 469$ to 829 MPa; η increased from $\eta = 0.061$ to 1.0; w_f increased from $w_f = 0.694$ to 1.795 μm; w_m decreased nonlinearly from $w_m = 0.676$ to 0.24 μm and remained unchanged. The CID stress was $\sigma_{cid} = 779$ MPa, the corresponding $u_{cod} = 1.438$ μm, $w_f = 1.678$ μm, and $w_m = 0.24$ μm.

- When $N = 10$, u_{cod} increased from $u_{cod} = 0.022$ to 1.58 μm with the increase in applied stress from $\sigma = 469$ to 829 MPa; η increased from $\eta = 0.065$ to 1.0; w_f increased from $w_f = 0.695$ to 1.804 μm; w_m decreased nonlinearly from $w_m = 0.673$ to 0.224 μm and remained unchanged. The CID stress was $\sigma_{cid} = 759$ MPa, the corresponding $u_{cod} = 1.416$ μm, $w_f = 1.64$ μm, and $w_m = 0.224$ μm.

Figure 4.10 shows the experimental and analytical predicted u_{cod} and related damage parameters with applied stress of SiC/SiC crack #4 for different cycles. When the cycle number increased from $N = 1$ to 10, the interface debonding state remained for partial debonding. u_{cod} and w_f both increased nonlinearly with tensile stress, and the w_m decreased nonlinearly with tensile stress. Under the same stress, u_{cod}, η, and w_f increased with the cycle number, and w_m decreased with applied cycles.

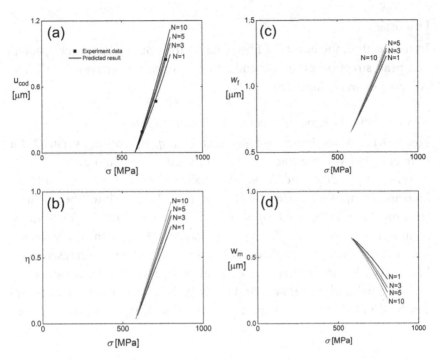

FIGURE 4.10 Experimental and predicted crack opening behavior of SiC/SiC crack #4 for N = 1, 3, 5, and 10: (a) u_{cod}, (b) η, (c) w_f, and (d) w_m.

- When N = 1, u_{cod} increased from u_{cod} = 0.0306 to 0.889 μm with the increase in applied stress from σ = 592 to 802 MPa; η increased from η = 0.064 to 0.736; w_f increased from w_f = 0.67 to 1.224 μm; and w_m decreased nonlinearly from w_m = 0.64 to 0.335 μm.

- When N = 3, u_{cod} increased from u_{cod} = 0.038 to 0.992 μm with the increase in applied stress from σ = 592 to 802 MPa; η increased from η = 0.071 to 0.816; w_f increased from w_f = 0.67 to 1.262 μm; and w_m decreased nonlinearly from w_m = 0.635 to 0.271 μm.

- When N = 5, u_{cod} increased from u_{cod} = 0.002 to 1.044 μm with the increase in applied stress from σ = 582 to 802 MPa; η increased from η = 0.037 to 0.858; w_f increased from w_f = 0.648 to 1.282 μm; and w_m decreased nonlinearly from w_m = 0.647 to 0.237 μm.

- When N = 10, u_{cod} increased from u_{cod} = 0.0048 to 1.122 μm with the increase in applied stress from σ = 582 to 802 MPa; η increased from η = 0.04 to 0.92; w_f increased from w_f = 0.649 to 1.312 μm; and w_m decreased nonlinearly from w_m = 0.645 to 0.189 μm.

Discussions

In this section, the effects of fiber volume fraction, matrix crack spacing, and peak stress on cyclic-dependent crack opening behavior in SiC/SiC composites were discussed.

Effect of Fiber Volume on Cyclic-Dependent CODs

Figure 4.11 shows the cycle-dependent u_{cod}, η, w_f, and w_m versus N for different fiber volume fractions (i.e., V_f = 20%, 25%, and 30%).

When V_f = 30% and 25%, PID occurred (i.e., $\eta < 1.0$), with the increase in applied cycles from N = 1 to 100; at low fiber volume fraction V_f = 20%, CID (η = 1.0) occurred at initial cyclic-fatigue loading (i.e., N_{CID} = 3). For V_f = 30% and 25%, u_{cod} and w_f increased nonlinearly with the applied cycle number, and w_m decreased non-linearly with the applied cycle number due to the increase in the interface debonding range. For V_f = 20%, u_{cod} and w_f versus N curves can be divided into two regions; before N = 3, u_{cod} and w_f increased

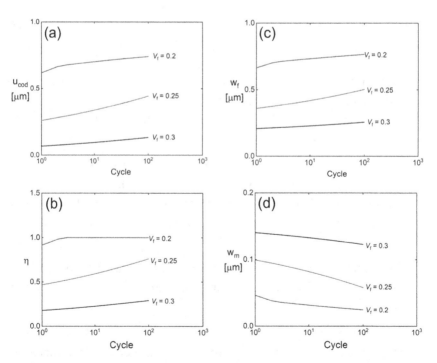

FIGURE 4.11 Analytical predicted crack opening parameters for V_f = 20%, 25%, and 30%: (a) u_{cod}, (b) η, (c) w_f, and (d) w_m.

nonlinearly with the applied cycle number, and when $N > 3$, u_{cod} and w_f increased linearly with the cycle number; w_m versus N curve can be divided into two regions; before $N = 3$, w_m decreased nonlinearly with the applied cycle number, and when $N > 3$, w_m decreased linearly with the cycle number.

- When $V_f = 20\%$, u_{cod} increased from $u_{cod} = 0.617$ μm at $N = 1$ to $u_{cod} = 0.739$ μm at $N = 10^2$ with η increased from $\eta = 91.6\%$ to 100% and w_f increased from $w_f = 0.664$ to 0.764 μm. w_m decreased nonlinearly from $w_m = 0.046$ to 0.024 μm. The cycle number for CID was $N_{CID} = 3$, the corresponding $u_{cod} = 0.676$ μm, $w_f = 0.721$ μm, and $w_m = 0.036$ μm.

- When $V_f = 25\%$, u_{cod} increased from $u_{cod} = 0.259$ μm at $N = 1$ to $u_{cod} = 0.444$ μm at $N = 10^2$ with η increased from $\eta = 46.8\%$ to 76.1% and w_f increased from $w_f = 0.358$ to 0.502 μm. w_m decreased nonlinearly from $w_m = 0.099$ to 0.057 μm.

- When $V_f = 30\%$, u_{cod} increased from $u_{cod} = 0.067$ μm at $N = 1$ to $u_{cod} = 0.132$ μm at $N = 10^2$ with η increased from $\eta = 17.9\%$ to 29.2% and w_f increased from $w_f = 0.207$ to 0.255 μm. w_m decreased nonlinearly from $w_m = 0.14$ to 0.122 μm.

Effect of Matrix Crack Spacing on Cyclic-Dependent CODs

Figure 4.12 shows cycle-dependent u_{cod}, η, w_f, and w_m versus N for different matrix crack spacing (i.e., $l_c = 300$, 400, and 500 μm). When the applied cycles increased from $N = 1$ to 10^2, η is less than 1.0 for $l_c = 300$, 400, and 500 μm. At the same applied cycle number, u_{cod} and η decreased with increasing matrix crack spacing, and w_f and w_m increased with matrix crack spacing.

- When $l_c = 300$ μm, u_{cod} increased from $u_{cod} = 0.093$ μm at $N = 1$ to $u_{cod} = 0.158$ μm at $N = 10^2$ with η increased from $\eta = 48.9\%$ to 79.9% and w_f increasing from $w_f = 0.125$ to 0.173 μm. w_m decreased nonlinearly from $w_m = 0.032$ to 0.015 μm.

- When $l_c = 400$ μm, u_{cod} increased from $u_{cod} = 0.088$ μm at $N = 1$ to $u_{cod} = 0.153$ μm at $N = 10^2$ with η increased from $\eta = 36.7\%$ to 59.9% and w_f increased from $w_f = 0.141$ to 0.189 μm. w_m decreased nonlinearly from $w_m = 0.053$ to 0.035 μm.

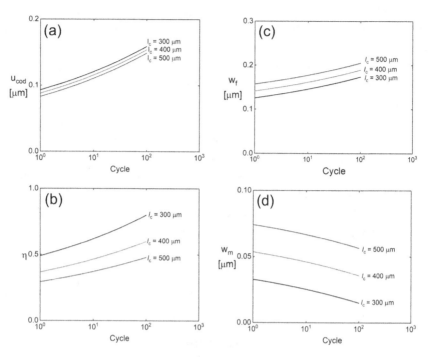

FIGURE 4.12   Analytical predicted crack opening parameters for l_c = 300, 400, and 500 μm: (a) u_{cod}, (b) η, (c) w_f, and (d) w_m.

- When l_c = 500 μm, u_{cod} increased from u_{cod} = 0.083 μm at N = 1 to u_{cod} = 0.148 μm at N = 10^2 with η increased from η = 29.4% to 47.9% and w_f increased from w_f = 0.157 to 0.205 μm. w_m decreased nonlinearly from w_m = 0.074 to 0.056 μm.

Effect of Peak Stress on Cyclic-Dependent CODs

Figure 4.13 shows cycle-dependent u_{cod}, η, w_f, and w_m versus N for different peak stresses (i.e., σ_{max} = 150, 200, and 250 MPa). Under σ_{max} = 150 and 200 MPa, η is less than 1.0, i.e., $\eta < 1.0$, when the cycle number increased from N = 1 to 10^2; and under σ_{max} = 250 MPa, η increased to 1.0, i.e., $\eta = 1.0$ at N_{CID} = 2. The cyclic-dependent IDR η depended on the peak stress and at high peak stress, the condition for CID occurred at the initial stage of fatigue loading. At the same applied cycle number, u_{cod}, η, and w_f increased with peak stress, and w_m decreased with peak stress, due to the increase in the interface debonding range. For σ_{max} = 150 and 200 MPa, u_{cod}, η, and w_f increased nonlinearly with the cycle number, and w_m decreased nonlinearly with the cycle number. For σ_{max} = 250, u_{cod}, η, and w_f versus N curves can be divided into two regions; before N = 2, u_{cod}, η, and w_f increased

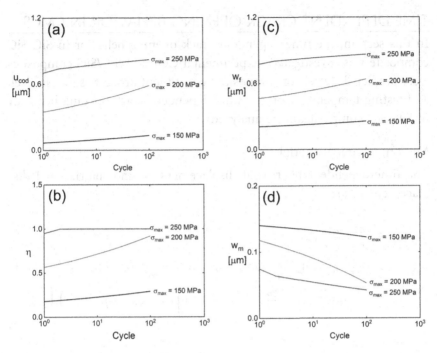

FIGURE 4.13 Analytical predicted crack opening parameters for σ_{max} = 150, 200, and 250 MPa: (a) u_{cod}, (b) η, (c) w_f, and (d) w_m.

nonlinearly with the cycle number, and when $N > 2$, u_{cod}, η, and w_f increased linearly with the cycle number; w_m versus N curve can be divided into two regions; before $N = 2$, w_m decreases nonlinearly with the cycle number, and when $N > 2$, w_m decreases linearly with the cycle number.

- When σ_{max} = 150 MPa, u_{cod} increased from u_{cod} = 0.0669 μm at $N = 1$ to u_{cod} = 0.132 μm at $N = 10^2$ with η increased from η = 17.9% to 29.2% and w_f increased from w_f = 0.207 to 0.255 μm. w_m decreased nonlinearly from w_m = 0.14 to 0.122 μm.

- When σ_{max} = 200 MPa, u_{cod} increased from u_{cod} = 0.346 μm at $N = 1$ to u_{cod} = 0.585 μm at $N = 10^2$ with η increased from η = 56.4% to 91.7% and w_f increased from w_f = 0.463 to 0.638 μm. w_m decreased nonlinearly from w_m = 0.117 to 0.053 μm.

- When σ_{max} = 250 MPa, u_{cod} increased from u_{cod} = 0.7 μm at $N = 1$ to u_{cod} = 0.819 μm at $N = 10^2$ with η increased from η = 94.9% to 100% and w_f increased from w_f = 0.775 to 0.862 μm. w_m decreased nonlinearly from w_m = 0.074 to 0.042 μm.

TIME-DEPENDENT CRACK OPENING BEHAVIOR IN CMCS

In this section, the time-dependent crack opening behavior in SiC/SiC composite was investigated. Experimental CODs of SiC/SiC composites were predicted. Effects of composite's constituent properties, stress level, and testing temperature on the time-dependent crack opening behavior in SiC/SiC composites were analyzed.

Micromechanical Model

The time-dependent fiber axial displacement w_f and matrix axial displacement w_f are

$$
\begin{aligned}
w_f(x) &= \int_x^{l_c/2} \frac{\sigma_f(x)}{E_f} dx \\
&= \frac{\sigma}{V_f E_f}(l_d - x) - \frac{\tau_f}{r_f E_f}(2\delta l_d - \delta^2 - x^2) - \frac{\tau_i}{r_f E_f}(l_d - \delta)^2 + \frac{\sigma_{fo}}{E_f}\left(\frac{l_c}{2} - l_d\right) \\
&\quad + \frac{r_f}{\rho E_f}\left[\frac{V_m}{V_f}\sigma_{mo} - \frac{2\tau_f}{r_f}\delta - \frac{2\tau_i}{r_f}(l_d - \delta)\right]\left[1 - \exp\left(-\rho\frac{l_c/2 - l_d}{r_f}\right)\right]
\end{aligned}
$$

$$(4.17)$$

$$
\begin{aligned}
w_m(x) &= \int_x^{l_c/2} \frac{\sigma_m(x)}{E_m} dx \\
&= \frac{V_f \tau_f}{r_f V_m E_m}(2\delta l_d - \delta^2 - x^2) + \frac{V_f \tau_i}{r_f V_m E_m}(l_d - \delta)^2 + \frac{\sigma_{mo}}{E_m}\left(\frac{l_c}{2} - l_d\right) \\
&\quad - \frac{r_f}{\rho E_m}\left[\sigma_{mo} - 2\frac{V_f \tau_f}{r_f V_m}\delta - 2\frac{V_f \tau_i}{r_f V_m}(l_d - \delta)\right]\left[1 - \exp\left(-\rho\frac{l_c/2 - l_d}{r_f}\right)\right]
\end{aligned}
$$

$$(4.18)$$

At the matrix crack plane, w_f and w_m can be obtained through Eqs. (4.17) and (4.18) by inserting $x = 0$, as follows:

$$
\begin{aligned}
w_f &= \frac{\sigma}{V_f E_f}l_d - \frac{\tau_f}{r_f E_f}(2\delta l_d - \delta^2) - \frac{\tau_i}{r_f E_f}(l_d - \delta)^2 + \frac{\sigma_{fo}}{E_f}\left(\frac{l_c}{2} - l_d\right) \\
&\quad + \frac{r_f}{\rho E_f}\left[\frac{V_m}{V_f}\sigma_{mo} - \frac{2\tau_f}{r_f}\delta - \frac{2\tau_i}{r_f}(l_d - \delta)\right]\left[1 - \exp\left(-\rho\frac{l_c/2 - l_d}{r_f}\right)\right]
\end{aligned}
$$

$$(4.19)$$

$$
\begin{aligned}
w_m &= \frac{V_f \tau_f}{r_f V_m E_m}(2\delta l_d - \delta^2) + \frac{V_f \tau_i}{r_f V_m E_m}(l_d - \delta)^2 + \frac{\sigma_{mo}}{E_m}\left(\frac{l_c}{2} - l_d\right) \\
&\quad - \frac{r_f}{\rho E_m}\left[\sigma_{mo} - 2\frac{V_f \tau_f}{r_f V_m}\delta - 2\frac{V_f \tau_i}{r_f V_m}(l_d - \delta)\right]\left[1 - \exp\left(-\rho\frac{l_c/2 - l_d}{r_f}\right)\right]
\end{aligned}
$$

$$(4.20)$$

Using Eqs. (4.19) and (4.20), the crack open displacement u_{cod} is

$$u_{cod} = \frac{2\sigma}{V_f E_f} l_d - \frac{2E_c \tau_f}{r_f V_m E_m E_f} l_d^2 (2\gamma - \gamma^2) - \frac{2E_c \tau_i}{r_f V_m E_m E_f} l_d^2 (1 - \gamma)^2$$
$$+ \frac{2r_f E_c}{\rho V_m E_m E_f} \left[\sigma_{mo} - 2\frac{\tau_f}{r_f}\delta - 2\frac{\tau_i}{r_f} l_d (1 - \gamma) \right] \left[1 - \exp\left(-\rho \frac{l_c}{2r_f} (1 - \eta) \right) \right]$$

$$(4.21)$$

where η and γ are the IDR and interface oxidation ratio, respectively, and defined by

$$\eta = \frac{2l_d}{l_c}, \ \gamma = \frac{\delta}{l_d} \qquad (4.22)$$

Experimental Comparisons

Figure 4.14 shows the experimental and predicted parameters (u_{cod}, η, w_f, and w_m) of SiC/SiC crack #1 for different durations (i.e., $t = 1$, 2, and 3 h)

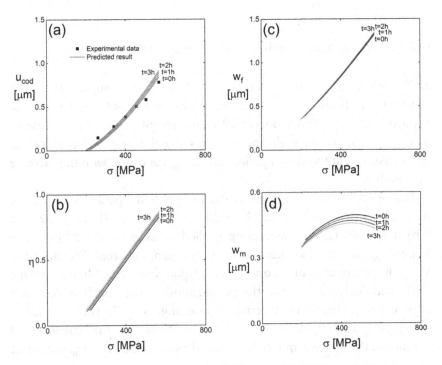

FIGURE 4.14 (a) u_{cod}, (b) η, (c) w_f, and (d) w_m of SiC/SiC crack #1.

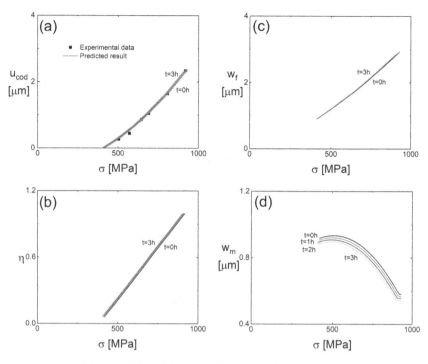

FIGURE 4.15   (a) u_{cod}, (b) η, (c) w_f, and (d) w_m of SiC/SiC crack #2.

at $T = 800°C$. u_{cod} and w_f increased nonlinearly with applied stress for different durations (i.e., $t = 1$, 2, and 3 h) at $T = 800°C$; at the same applied stress, u_{cod} and w_f increased with duration due to the increase in η. w_m first increased with applied stress to the peak value and then decreased; and at the same applied stress, w_m decreased with the increase in duration.

Figure 4.15 shows the experimental and predicted parameters (u_{cod}, η, w_f, and w_m) of SiC/SiC crack #2 for different durations (i.e., $t = 1$, 2, and 3 h) at $T = 800°C$. With increasing applied stress, η increased to $\eta = 1.0$, indicating the interface debonding state changing from PID to CID. With the increase in duration, the corresponding stress for the damage CID state decreased, due to the propagation of the interface oxidation region. The change of the interface debonding state affects the evolution of u_{cod}, w_f, and w_m with the increase in stress. For the damage CID condition, the increase in rate of u_{cod} and w_f decreased and w_m remained constant.

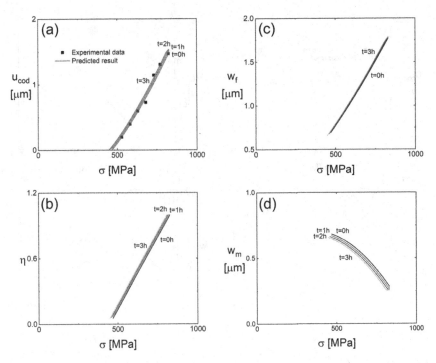

FIGURE 4.16 (a) u_{cod}, (b) η, (c) w_f, and (d) w_m of crack #3.

Figure 4.16 shows the experimental and predicted parameters (u_{cod}, η, w_f, and w_m) of SiC/SiC crack #3 for different durations (i.e., $t = 1, 2$, and 3 h) at $T = 800°C$. With the increase in duration from $t = 1$ to 3 h, the stress for the damage CID state decreased from $\sigma_{cid}=829$ to 809 MPa. u_{cod} and w_f increased nonlinearly with applied stress for the damage PID condition and then linearly with applied stress for the damage CID condition. w_m decreased nonlinearly with applied for PID and remained constant for CID.

Figure 4.17 shows the experimental and predicted parameters (u_{cod}, η, w_f, and w_m) of SiC/SiC crack #4 for different durations (i.e., $t = 1, 2$, and 3 h) at $T = 800°C$. With the increase in applied stress and duration, the interface debonding state remained PID, i.e., $\eta < 1.0$. u_{cod} and w_f increased nonlinearly with applied stress, and w_m decreased non-linearly with applied stress. With the increase in duration, the values of u_{cod} and w_f increased and the value of w_m decreased at the same applied stress.

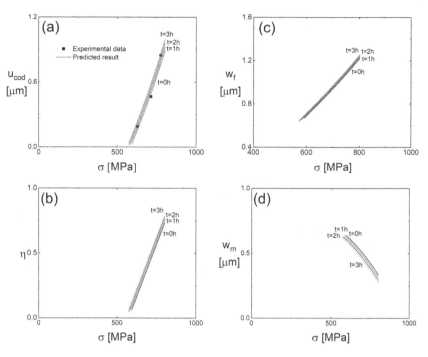

FIGURE 4.17   (a) u_{cod}, (b) η, (c) w_f, and (d) w_m of SiC/SiC crack #4.

Discussions

For the crack opening behavior in SiC/SiC composites, the composite's constitutive properties and stress level affect the composite's internal damage. In this section, the effects of composite's constituent properties, stress level, and temperature on the time-dependent crack opening behavior in SiC/SiC composites were analyzed.

Effect of Constitutive Properties on Time-Dependent CODs

Figure 4.18 shows the effect of interface shear stress (i.e., τ_i = 2, 4, and 6 MPa) on the time-dependent crack opening behavior in SiC/SiC composites at T = 800°C. When τ_i increased from τ_i = 2 to 6 MPa, the interface debonding length decreased and the value of η also decreased, leading to the decrease in u_{cod} and w_f and the increase in w_m. When τ_i increased, the interface debonding length decreased, the load transfer capacity in the slip region increased, and more external load was carried by the matrix, leading to the decrease in u_{cod}.

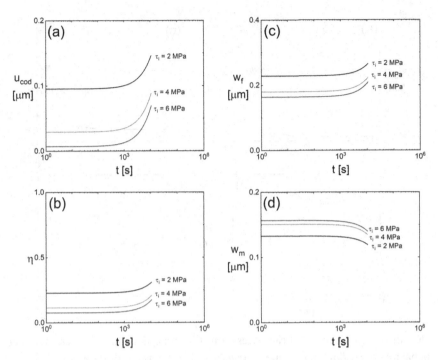

FIGURE 4.18 Effect of τ_i on the time-dependent crack opening behavior in SiC/ SiC composites: (a) u_{cod}, (b) η, (c) w_f, and (d) w_m.

Effect of Stress Level on Time-Dependent CODs

Figure 4.19 shows the effect of stress level (i.e., σ_{max} = 150, 250, and 350 MPa) on time-dependent crack opening behavior in SiC/SiC composites at T = 800°C. Under σ_{max} = 350 MPa, CID occurred upon initial loading, i.e., η = 1.0. Under σ_{max} = 250 MPa, η increased with duration and approached η = 1.0. Under σ_{max} = 150 MPa, the interface debonding state remained PID with the increase in duration. The values of u_{cod} and w_f both increased with applied stress and the value of w_m decreased with applied stress, mainly due to the increase in η. For the CID condition, the propagation of the interface oxidation region in the debonding region increased the values of u_{cod} and w_f and decreased the value of w_m.

Effect of Temperature on Time-Dependent CODs

Figure 4.20 shows the effect of temperature (i.e., T = 800, 900, and 1000°C) on the time-dependent crack opening behavior of SiC/SiC composites under σ_{max} = 200 MPa. u_{cod} and w_f both increased with temperature from T = 800 to 1000°C due to the increase in η. w_m

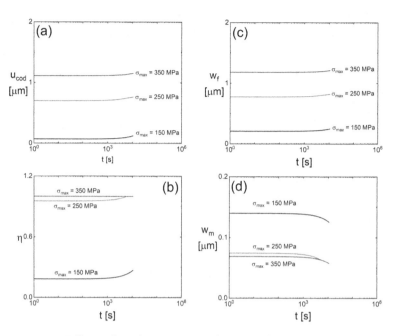

FIGURE 4.19 Effect of peak stress on the time-dependent crack opening behavior in SiC/SiC composites: (a) u_{cod}, (b) η, (c) w_f, and (d) w_m.

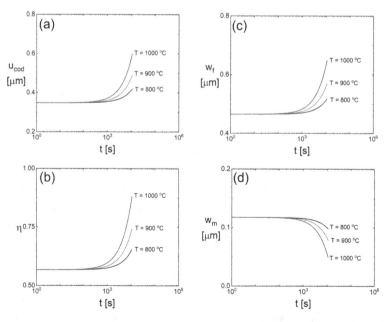

FIGURE 4.20 Effect of temperature on the time-dependent crack opening behavior in SiC/SiC composites: (a) u_{cod}, (b) η, (c) w_f, and (d) w_m.

decreased with temperature from $T = 800$ to $1000°C$ due to the increase in η. With the increase in temperature, the interface oxidation rate increased, leading to the increase in η and the low load transfer between the fiber and the matrix.

SUMMARY AND CONCLUSIONS

In this chapter, the stress-dependent, cyclic-dependent, and time-dependent crack opening behavior in SiC/SiC composites were investigated. Multiple damage mechanisms of interface debonding, interface wear, and interface oxidation were considered. At the matrix cracking plane, the COD, IDR, and related fiber and matrix axial displacements were derived. Experimental CODs of SiC/SiC composite were predicted. Effects of composite's constituent properties, stress level, cycle number, and temperature on crack opening behavior in SiC/SiC composites were also analyzed.

REFERENCES

1. Naslain R. SiC-matrix composites: Nonbrittle ceramics for thermostructural application. *Int. J. Appl. Ceram. Technol.* 2005; 2:75–84.
2. DiCarlo JA, Yun HM, Morscher GN, Bhatt RT. SiC/SiC composites for 1200°C and above. In: Bansal N.P. (ed.) *Handbook of ceramic composites*. Springer, Boston, MA. 2005.
3. Li LB. *Damage, fracture and fatigue of ceramic-matrix composites*. Springer Nature, Singapore. 2018.
4. Barsoum MW, Kangutkar P, Wang ASD. Matrix crack initiation in ceramic matrix composites Part I: Experiments and test results. *Compos. Sci. Technol.* 1992; 44:257–269.
5. Danchavijit S, Shetty DK. Matrix cracking in ceramic-matrix composites. *J. Am. Ceram. Soc.* 1993; 76:2497–2504.
6. Liu S, Zhang L, Yin X, Liu Y, Cheng L. Proportional limit stress and residual thermal stress of 3D SiC/SiC composite. *J. Mater. Sci. Technol.* 2014; 30:959–964.
7. Filipuzzi L, Campus G, Naslain R. Oxidation mechanisms and kinetics of 1D-SiC/C/SiC composite materials: I, An experimental approach. *J. Am. Ceram. Soc.* 1994; 77:459–466.
8. Li LB. A micromechanical crack opening displacement model for fiber-reinforced ceramic-matrix composites considering matrix fragmentation. *Theor. Appl. Fract. Mech.* 2021; 112:102875.
9. Li LB. Interface wear effects in ceramic composite crack opening. *J. Compos. Mater.* 2022; 56:3371–3384.

10. Li LB. Micromechanical modeling of time-dependent crack opening behavior in SiC/SiC composites. Proceedings of ASME Turbo Expo 2022 Turbomachinery Technical Conference and Exposition GT2022. June 13–17, 2022, Rotterdam, The Netherlands.

11. Forna-Kreutzer JP, Ell J, Barnard H, Pizada T, Ritchie R, Liu D. Full-field characterization of oxide-oxide ceramic-matrix composites using X-ray computed micro-tomography and digital volume correlation under load at high temperatures. *Mater. Design* 2021; 208:109899.

12. Jordan SP, Newton CD, Nicholson PI, Jeffs SP, Gale L, Bache MR. Matrix crack networks in SiC/SiC composites: In-situ characterization and metrics. ASME Turbo Expo 2021: Turbomachinery Technical Conference and Exposition, GT 2021, June 7–11, 2021.

13. Chen Y, Gelebart L, Chateau C, Bornert M, King A, Aimedieu P, Sauder C. 3D detection and quantitative characterization of cracks in a ceramic matrix composite tube using X-ray computed tomography. *Exp. Mech.* 2020; 60:409–424.

14. Santhosh U, Ahmad J, Ojard G, Gowayed Y. A synergistic model of stress and oxidation induced damage and failure in silicon carbide-based ceramic matrix composites. *J. Am. Ceram. Soc.* 2021; 104:4163–4182.

15. Harrison S, Schneiter J, Pegna J, Vaaler E, Goduguchinta R, Williams K. High-temperature performance of next-generation silicon carbide fibers for CMCs. *Mater. Perform. Charact.* 2021; 10: 207–223.

16. Chateau C, Gelebart L, Bornert M, Crepin J, Caldemaison D, Sauder C. Modeling of damage in unidirectional ceramic matrix composites and multi-scale experimental validation on third generation SiC/SiC mini-composites. *J. Mech. Phys. Solids* 2014; 63:298–319.

17. Sevener KM, Tracy JM, Chen Z, Kiser JD, Daly S. Crack opening behavior in ceramic matrix composites. *J. Am. Ceram. Soc.* 2017; 100:4734–4747.

18. Morscher GN, Singh M, Kiser JD, Freedman M, Bhatt R. Modeling stress-dependent matrix cracking and stress-strain behavior in 2D woven SiC fiber reinforced CVI SiC composites. *Compos. Sci. Technol.* 2007; 67:1009–1017.

19. Simon C, Rebillat F, Herb V, Camus G. Monitoring damage evolution of $SiC_f/[Si\text{-}B\text{-}C]_m$ composites using electrical resistivity: Crack density-based electromechanical modeling. *Acta Mater.* 2017; 124:579–587.

CHAPTER 5

High-Temperature Cracking Closure Behavior in Ceramic-Matrix Composites

INTRODUCTION

For the design of ceramic-matrix composite (CMC) components, it is necessary to determine the design stress. For SiC/SiC composites, the proportional limit stress (PLS) is usually used as the design stress [1,2]. However, below the PLS, matrix cracking may also appear, and the crack opening stress and crack closure stress (CCS) may be much lower than the PLS. For some CMCs, the matrix cracks can contact during unloading, even when the matrix is subject to residual tension, which is defined as matrix cracking closure [3,4], as shown in Figure 5.1. Such

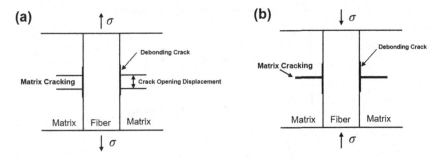

FIGURE 5.1 Schematic of matrix crack (a) opening and (b) closure in fiber-reinforced CMCs.

DOI: 10.1201/9781032638508-5

behavior arises due to the occurrence of lateral grain-to-grain displacements as the matrix cracks form [5,6]. The thermal expansion coefficient of the fiber and the matrix changes with temperature, indicating the change in thermal residual stress (TRS) of CMCs with temperature [7–9]. To ensure the reliability and durability of CMC components, it is necessary to determine the value of CCS for different CMCs [10,11].

STRESS-DEPENDENT CRACK CLOSURE BEHAVIOR OF CMCS

Figure 5.2 shows the experimental and predicted matrix cracking density λ_m versus loading/unloading applied stress σ curves in SiC/SiC composites [11]. Upon loading, the matrix cracking density λ_m increased slowly at low applied stress, then increased rapidly to the peak value, and finally approached saturation. Upon unloading, the matrix cracking density λ_m decreased slowly with applied stress to the original value. At an intermediate stress level, the matrix cracking density λ_m upon unloading was much higher than that upon loading. The difference between the loading and unloading matrix cracking density λ_m curve affects the proportionality of the interface debonding in matrix crack spacing. Under the high applied stress, the matrix cracking density λ_m is high and the matrix crack spacing is short, and the interface debonding length may occupy the entire matrix crack; however, under the low applied stress, the matrix cracking density λ_m is low and the matrix crack spacing is long, and the interface debonding length only occupies partial matrix crack spacing.

Upon loading/unloading tensile, the difference of matrix crack density λ_m curve affects the mechanical hysteresis loops. Figure 5.3 shows the

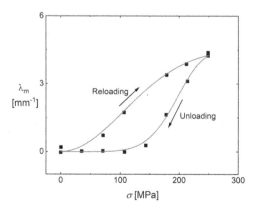

FIGURE 5.2   Experimental and predicted evolution of matrix cracking density upon loading and reloading.

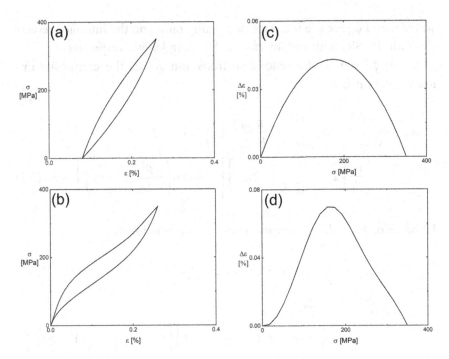

FIGURE 5.3 (a) Hysteresis loop of non-closure crack; (b) hysteresis loop of closure crack; (c) strain difference of non-closure crack; and (d) strain difference of closure crack in SiC/SiC composites.

loading/unloading mechanical hysteresis loops and the strain difference between the loading and the unloading process. Due to the evolution of matrix cracking density and the difference in matrix cracking density upon loading/unloading, the unloading and reloading strain coincide with each other before the valley stress, and the hysteresis loop width is much larger than that for the case without considering matrix cracking closure. In this section, the mechanical hysteresis analysis is adopted to analyze the matrix crack closure behavior in SiC/SiC composites.

Micromechanical Model

Upon unloading, the composite strain is equivalent to the undamaged fiber strain and is given by

$$
\begin{aligned}
\varepsilon_u = {} & \frac{\sigma_u}{V_f E_f}\eta + 2\frac{\tau_i}{E_f}\frac{l_d}{r_f}\eta\phi_u^2 - \frac{\tau_i}{E_f}\frac{l_d}{r_f}\eta\left(2\phi_u - 1\right)^2 + \frac{\sigma_{fo}}{E_f}\left(1 - \eta\right) \\
& + \frac{1}{\rho E_f}\left[2\sigma_{mo}\frac{V_m}{V_f}\frac{r_f}{l_c} - 2\tau_i\eta\left(1 - 2\phi_u\right)\right]\left[1 - \exp\left(-\frac{\rho l_c}{2r_f}\left(1 - \eta\right)\right)\right] - \varepsilon_t
\end{aligned}
\tag{5.1}
$$

where η and ϕ_u are the interface debonding ratio and the interface reverse slip ratio (IRSR) with the interface debonding length, respectively.

When $\phi_u = 1$ at the unloading transition stress, the composite unloading strain is

$$
\begin{aligned}
\varepsilon_u =\ & \frac{\sigma_u}{V_f E_f}\eta + 2\frac{\tau_i}{E_f}\frac{l_d}{r_f}\eta - \frac{\tau_i}{E_f}\frac{l_d}{r_f}\eta + \frac{\sigma_{fo}}{E_f}(1 - \eta) \\
& + \frac{1}{\rho E_f}\left[2\sigma_{mo}\frac{V_m}{V_f}\frac{r_f}{l_c} + 2\tau_i\eta\right]\left[1 - \exp\left(-\frac{\rho l_c}{2r_f}(1 - \eta)\right)\right] - \varepsilon_t \quad (5.2)
\end{aligned}
$$

Upon reloading, the composite stress–strain relationship is

$$
\begin{aligned}
\varepsilon_r =\ & \frac{\sigma_r}{V_f E_f}\eta - 2\frac{\tau_i}{E_f}\frac{l_d}{r_f}\eta\phi_r^2 + 2\frac{\tau_i}{E_f}\frac{l_d}{r_f}\eta(1 - 2\phi_r)^2 \\
& - \frac{\tau_i}{E_f}\frac{l_d}{r_f}\eta(1 - 2\phi_r)^2 + \frac{\sigma_{fo}}{E_f}(1 - \eta) \\
& + \frac{2}{\rho E_f}\left[\sigma_{mo}\frac{V_m}{V_f}\frac{r_f}{l_c} - \tau_i\eta(1 - 2\eta + 2\phi_r)\right]\left[1 - \exp\left(-\frac{\rho}{2}\frac{l_c}{r_f}(1 - \eta)\right)\right] \\
& - \varepsilon_t
\end{aligned}
$$

$$(5.3)$$

where ϕ_r is the interface new slip ratio with the interface debonding length.

When $\phi_r = 1$ at the reloading transition stress, the composite reloading strain is

$$
\begin{aligned}
\varepsilon_r =\ & \frac{\sigma_r}{V_f E_f}\eta - 2\frac{\tau_i}{E_f}\frac{l_d}{r_f}\eta + 2\frac{\tau_i}{E_f}\frac{l_d}{r_f}\eta - \frac{\tau_i}{E_f}\frac{l_d}{r_f}\eta + \frac{\sigma_{fo}}{E_f}(1 - \eta) \\
& + \frac{2}{\rho E_f}\left[\sigma_{mo}\frac{V_m}{V_f}\frac{r_f}{l_c} - \tau_i\eta(3 - 2\eta)\right]\left[1 - \exp\left(-\frac{\rho}{2}\frac{l_c}{r_f}(1 - \eta)\right)\right] \quad (5.4) \\
& - \varepsilon_t
\end{aligned}
$$

The strain difference between loading and unloading $\Delta\varepsilon$ is

$$
\Delta\varepsilon(\sigma) = \varepsilon_u(\sigma_u) - \varepsilon_r(\sigma_r) \quad (5.5)
$$

Upon unloading and reloading, the damage parameters of ϕ_u and ϕ_r change with the decrease or increase in applied stress due to change in interface slip and matrix cracking density. Evolution of ϕ_u and ϕ_r versus loading/unloading stress can be used to determine the CCS of SiC/SiC composites. The CCS is determined from the ϕ_u versus σ graph and comes from the point at which ϕ_u becomes zero, i.e., $\phi_u \leq 10^{-2}$.

Experimental Comparisons

Sauder et al. [12], Morscher and Gordon [13], and Smith [14] performed experimental investigations on the cyclic loading/unloading tensile behavior of 1D mini and unidirectional, 2D cross-ply, and plain-woven SiC/SiC composites. In the present analysis, experimental matrix cracking closure stress in different SiC/SiC composites is predicted using the developed hysteresis-based identification approach.

Cracking Closure Stress of Mini-SiC/SiC Composites

Figures 5.4–5.6 show the experimental and predicted mechanical hysteresis loops and damage parameter ϕ_u of 1D mini-SiC/SiC composites

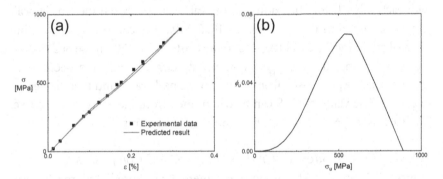

FIGURE 5.4 (a) Hysteresis loops and (b) ϕ_u versus σ_u curve under $\sigma_{max} =$ 890 MPa of mini-SiC/SiC composites.

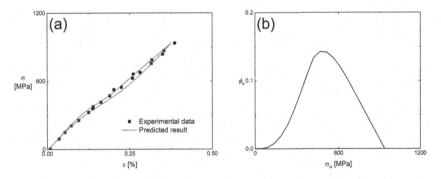

FIGURE 5.5 (a) Hysteresis loops and (b) ϕ_u versus σ_u curve under $\sigma_{max} =$ 935 MPa of mini-SiC/SiC composites.

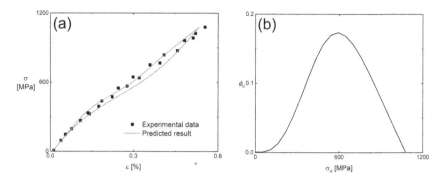

FIGURE 5.6 (a) Hysteresis loops and (b) ϕ_u versus σ_u curve under σ_{max} = 1078 MPa of mini-SiC/SiC composites.

under σ_{max} = 890, 935, and 1078 MPa. As the matrix cracking density changes with the increase in applied stress upon loading and decrease in applied stress upon unloading, the unloading and reloading strain coincide with each other before unloading to the valley stress. When the tensile peak stress increases from σ_{max} = 890 to 1078 MPa, the composite's unloading residual strain increases a little, due to the existence of TRS in the SiC matrix. Upon unloading, the damage parameter ϕ_u changes with the decrease in applied stress. ϕ_u increased from ϕ_u = 0 to the peak value and then decreased to ϕ_u = 0. The value of CCS can be determined from the ϕ_u versus σ_u curve. The CCS is σ_{ccs} = 108 MPa for 1D mini-SiC/SiC composites.

Cracking Closure Stress of Unidirectional SiC/SiC Composites
Figures 5.7 and 5.8 show the experimental and predicted mechanical hysteresis loops and damage parameter ϕ_u of unidirectional SiC/SiC

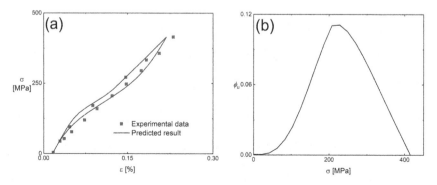

FIGURE 5.7 (a) Hysteresis loops and (b) ϕ_u versus σ_u curve under σ_{max} = 415 MPa of unidirectional SiC/SiC composites.

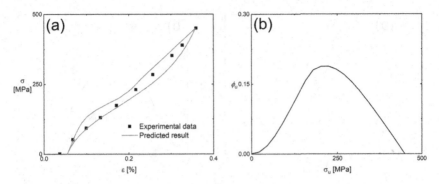

FIGURE 5.8 (a) Hysteresis loops and (b) ϕ_u versus σ_u curve under σ_{max} = 450 MPa of unidirectional SiC/SiC composites.

composites under σ_{max} = 415 and 450 MPa. The value of CCS can be determined from the evolution of ϕ_u versus σ_u curves for different peak stresses. Under σ_{max} = 415 MPa, the CCS is σ_{ccs} = 42 MPa; under σ_{max} = 450 MPa, the CCS is σ_{ccs} = 20 MPa. For unidirectional SiC/SiC composites, when the tensile peak stress increases, the CCS decreases.

Cracking Closure Stress of Cross-Ply SiC/SiC Composites
For cross-ply SiC/SiC composites, transverse cracks in the 90° plies occur under tensile loading. However, interface debonding or slip does not exist in these transverse cracks. Matrix cracks in the 0° plies with interface debonding and slip mainly contribute to the cracking opening and closure.

Figures 5.9 and 5.10 show the experimental and predicted hysteresis loops and damage parameter ϕ_u of cross-ply SiC/SiC composites under

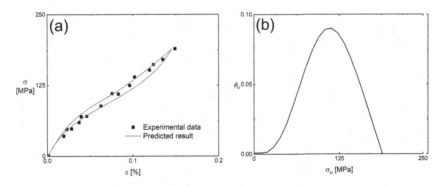

FIGURE 5.9 (a) Hysteresis loops and (b) ϕ_u versus σ_u curve under σ_{max} = 189 MPa of cross-ply SiC/SiC composites.

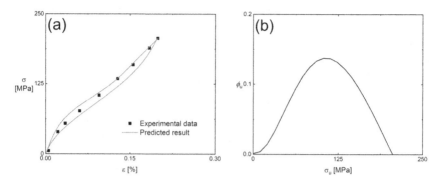

FIGURE 5.10 (a) Hysteresis loops and (b) ϕ_u versus σ_u curve under σ_{max} = 206 MPa of cross-ply SiC/SiC composites.

σ_{max} = 189 and 206 MPa. The value of CCS can be determined from the ϕ_u versus σ_u curves for different peak stresses. Under σ_{max} = 189 MPa, the CCS is σ_{ccs} = 20 MPa; under σ_{max} = 206 MPa, the CCS is σ_{ccs} = 10 MPa. For cross-ply SiC/SiC composites, when the tensile peak stress increases, the CCS decreases.

Cracking Closure Stress of 2D Woven SiC/SiC Composites
Figures 5.11–5.13 show the experimental and predicted mechanical hysteresis loops and damage parameter ϕ_u of 2D plain-woven SiC/SiC composites under σ_{max} = 312, 360, and 410 MPa. The values of CCS can be determined from the ϕ_u versus σ_u curves for different peak stresses. When the tensile peak stress increases from σ_{max} = 312 to 410 MPa, the CCS also remains the same, i.e., σ_{ccs} = 16 MPa. When the tensile peak stress increases, the interface debonding and slip state at CCS remain

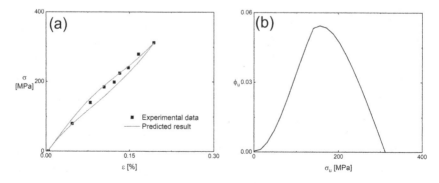

FIGURE 5.11 (a) Hysteresis loops and (b) ϕ_u versus σ_u curve under σ_{max} = 312 MPa of 2D plain-woven SiC/SiC composites.

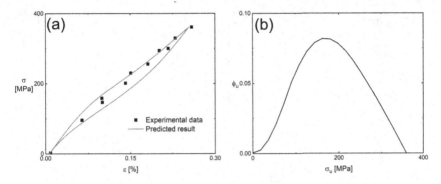

FIGURE 5.12   (a) Hysteresis loops and (b) ϕ_u versus σ_u curve under $\sigma_{max} = 360$ MPa of 2D plain-woven SiC/SiC composites.

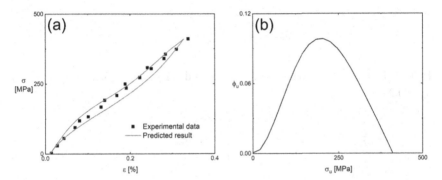

FIGURE 5.13   (a) Hysteresis loops and (b) ϕ_u versus σ_u curve under $\sigma_{max} = 410$ MPa of 2D plain-woven SiC/SiC composites.

unchanged, which may be attributed to the thermal residual compressive stress in the SiC matrix.

Discussions

In this section, hysteresis analysis was adopted to determine the value of CCS. Interface debonding and slip state affect the CCS in CMCs. Effects of fiber volume fraction and interface properties on the CCS and related interface damage were also discussed.

Effect of Fiber Volume on Crack Closure of CMCs

Figures 5.14–5.16 show the mechanical hysteresis loops, strain differences $\Delta\varepsilon$ between unloading and reloading strain, and ϕ_u of SiC/SiC composites for different V_f (i.e., $V_f = 0.3$, 0.35, and 0.4). When V_f

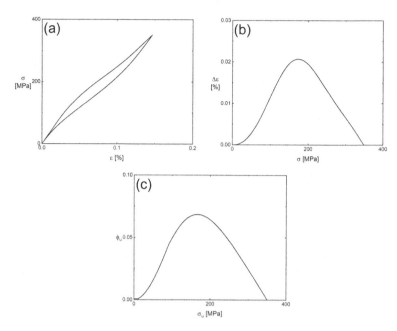

FIGURE 5.14   (a) Hysteresis loops, (b) $\Delta\varepsilon$, and (c) ϕ_u of SiC/SiC composites for $V_f = 0.3$.

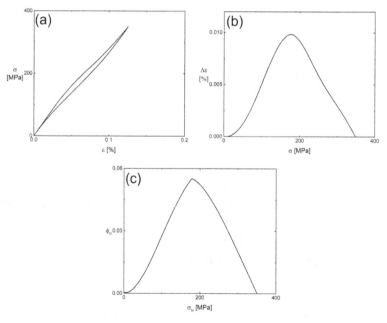

FIGURE 5.15   (a) Hysteresis loops, (b) $\Delta\varepsilon$, and (c) ϕ_u of SiC/SiC composites for $V_f = 0.35$.

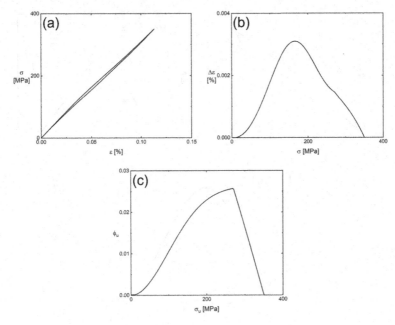

FIGURE 5.16   (a) Hysteresis loops, (b) $\Delta\varepsilon$, and (c) ϕ_u of SiC/SiC composites for $V_f = 0.4$.

increased from $V_f = 0.3$ to 0.4, the strain difference $\Delta\varepsilon$ between the unloading and reloading strain decreased due to the decrease in the interface slip range. Under low V_f, i.e., $V_f = 0.3$, the unloading stress corresponding to the peak ϕ_u was much lower than that of high V_f, i.e., $V_f = 0.35$ and 0.4, and the CCS increased with V_f.

Effect of Interface Shear Stress on Crack Closure of CMCs

Figures 5.17–5.19 show the mechanical hysteresis loops, strain difference $\Delta\varepsilon$ between unloading and reloading strain, and ϕ_u of SiC/SiC composites for different τ_i (i.e., $\tau_i = 10$, 20, and 30 MPa). When τ_i increased from $\tau_i = 10$ to 30 MPa, the strain difference $\Delta\varepsilon$ between the unloading and reloading strain decreased due to the decrease of the interface slip range, and the peak values for ϕ_u decreased. When the value τ_i increased, the frictional resistance for interface debonding and slip increased. Upon unloading, the value of the CCS increased with the increase in τ_i.

Effect of TRS on Crack Closure of CMCs

The crack closure behavior in CMCs depends on the TRS which develops due to the mismatch of elastic constants and coefficients of thermal

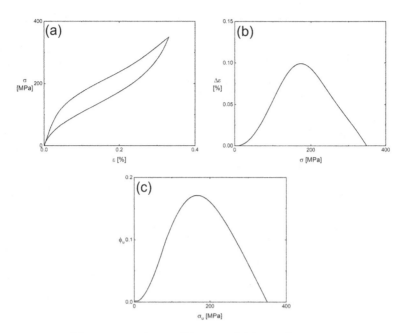

FIGURE 5.17   (a) Hysteresis loops, (b) $\Delta\varepsilon$, and (c) ϕ_u of SiC/SiC composites for $\tau_i = 10$ MPa.

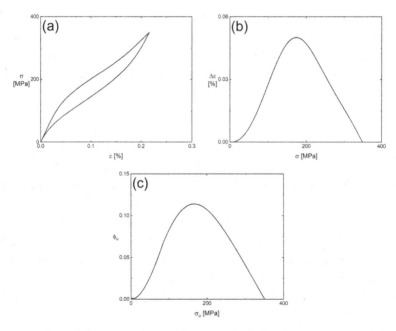

FIGURE 5.18   (a) Hysteresis loops, (b) $\Delta\varepsilon$, and (c) ϕ_u of SiC/SiC composites for $\tau_i = 20$ MPa.

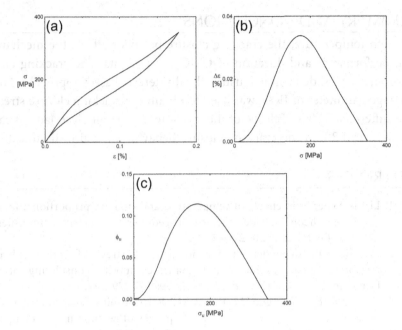

FIGURE 5.19 (a) Hysteresis loops, (b) $\Delta\varepsilon$, and (c) ϕ_u of SiC/SiC composites for $\tau_i = 30$ MPa.

expansion (CTEs) between the fiber and the matrix. For the composite with ideal interfacial bonds, the axial residual stress in a non-cracked matrix is [9]

$$\sigma_{\text{trs\_axial}}^{\text{matrix}} = E_m \frac{\chi V_f E_f}{\chi V_f E_f + V_m E_m}(\alpha_f - \alpha_m)(T_o - T_p) \qquad (5.6)$$

where χ is the non-unidirectional reinforcement correction factor, α_f and α_m denote the fiber and matrix CTEs, respectively, and T_o and T_p denote the testing and processing temperatures. For 2D woven C/SiC composites, the axial TRS in non-cracked matrix determined by Eq. (5.6) was about 282 MPa [9]. The residual tensile stress in the SiC matrix causes the opening of microcracks in the composite once cooled down from the process temperature to room temperature. Under tensile loading, with increasing tensile stress, new matrix cracking and interface debonding would relieve the TRS in the matrix. Using the common intersection point method, the axial TRS in the cracking matrix was about 130 MPa [9]. At elevated temperature, the axial TRS in the matrix would decrease as shown in Eq. (5.6), and the matrix cracking closure stress would also decrease.

SUMMARY AND CONCLUSIONS

At high temperature, the cracking closure behavior affects the mechanical performance and duration of CMC components. The cracking closure stress was determined using the hysteresis-based approach. The damage parameter of IRSR was used to obtain the cracking closure stress for different CMCs. Effects of fiber volume fraction, interface shear stress, and TRS on the cracking closure behavior were also discussed.

REFERENCES

1. Li LB. Synergistic effects of temperature and time on proportional limit stress of silicon carbide fiber-reinforced ceramic-matrix composites. *Compos. Interfaces* 2020; 27:341–353.
2. Li LB. Time-dependent proportional limit stress of carbon fiber-reinforced silicon carbide ceramic-matrix composites considering interface oxidation. *J. Ceram. Soc. Japan* 2019; 127:279–287.
3. Vagaggini E, Domergue JM, Evans AG. Relationships between hysteresis measurements and the constituent properties of ceramic matrix composites: I, Theory. *J. Am. Ceram. Soc.* 1995; 78:2709–2720.
4. Reynaud P, Dalmaz A, Tallaron C, Rouby D, Fantozzi G. Apparent stiffening of ceramic-matrix composites induced by cyclic fatigue. *J. Eur. Ceram. Soc.* 1998; 18:1827–1833.
5. Kotil T, Holmes JW, Comninou M. Origin of hysteresis observed during fatigue of ceramic-matrix composites. *J. Am. Ceram. Soc.* 1990; 73:1879–1883.
6. Steen M. Tensile mastercurve of ceramic matrix composites: Significance and implications for modeling. *Mater. Sci. Eng. A* 1998; 250:241–248.
7. Camus G. Guillaumat L, Baste S. Development of damage in a 2D woven C/SiC composite under mechanical loading 1 mechanical characterization. *Compos. Sci. Technol.* 1996; 56:1363–1372.
8. Dassios KG, Aggelis DG, Kordatos EZ, Matikas TE. Cyclic loading of a SiC-fiber reinforced ceramic matrix composite reveals damage mechanisms and thermal residual stress state. *Compos. Part A* 2013; 44:10–113.
9. Mei H. Measurement and calculation of thermal residual stress in fiber reinforced ceramic matrix composites. *Compos. Sci. Technol.* 2008; 68:3285–3292.
10. Li LB. Hysteresis-based identification approach for crack opening and closure stress in SiC/SiC fiber-reinforced ceramic-matrix composites. *Int. J. Fatigue* 2022; 162:106945.
11. Li LB. A micromechanical loading/unloading constitutive model of fiber-reinforced ceramic-matrix composites considering matrix crack closure. *Fatigue Fract. Eng. Mater. Struct.* 2021; 44:2389–2411.

12. Sauder C, Brusson A, Lamon J. Influence of interface characteristics on the mechanical properties of Hi-Nicalon type-S or Tyranno-SA3 fiber-reinforced SiC/SiC minicomposites. *Int. J. Appl. Ceram. Technol.* 2010; 7:291–303.
13. Morscher GN, Gordon NA. Acoustic emission and electrical resistance in SiC-based laminate ceramic composites tested under tensile loading. *J. Eur. Ceram. Soc.* 2017; 37:3861–3872.
14. Smith CE. Electrical resistance changes of melt infiltrated SiC/SiC subjected to long-term tensile loading at elevated temperature. PhD thesis. University of Akron, USA, 2016.

Index

Note: Page numbers in *italics* refer to figures.